T0149304

Free Fishes

Durime P. Zherka

authorHOUSE®

AuthorHouse™
1663 Liberty Drive
Bloomington, IN 47403
www.authorhouse.com
Phone: 1 (800) 839-8640

Published by AuthorHouse   02/26/2016

ISBN: 978-1-5049-8187-3 (sc)
ISBN: 978-1-5049-8186-6 (e)

Library of Congress Control Number: 2016903207

Print information available on the last page.

This book is printed on acid-free paper.

Contents

Fishes

Chapter 1

School Fishes

One very beautiful sunny day, while the rays were going through the blue clear water one beautiful fish was seeing with wondering one big crew two colors white –yellow fishes, and it understood that those were school's fishes. How beautiful is this crew was thinking the fish, all are happy are swimming together, looks that they are happy with their school. When the school crew fishes were swimming close to this single fish they gave greeting to it and splashed water all around. Really this school crew of fishes were very happy and they have decided to swim more longer around and to play with each other.

The single fish that was called by others Tropical fish that was proud for his strip yellow and white was thinking to go and to discus with his close friend Yellow Tank fish about school. Tropical fish knows that he was very young for school but he likes this school crew fishes, so he thought why he did not start school yet. Tropical fish has heard by his parents that before to go to school must to take some classes preschool and to prepare itself for longer swimming.

At that time of its thinking suddenly came close to it with full happiness the Seggelflossen doctor. Tropical fish without breath asked: I see you happy but I think you are going to see any sick fish?

No, no, no answered Zebrasoma that really is its nickname, is not any fish sick but I am going to one seminar about vaccination's new center. This center is new one and will use material by some turtled plasm for that vaccine that is done experiment and the result is so good with big success.

When is opened this center asked Tropical fish?

Is not open yet but we are preparing it, and very soon we will open that center.

But you dear Tropical why are alone here, you are very young to be alone in this area far away from your parents?

I was swimming around but I saw school's fishes, and I got shocked by their happiness and I did not understood how I was swimming without direction while I was seeing them that were laughing and playing with each other.

Okay now go to your parents!

No Dear Zabrasoma I decided to go to meet one friend of mine and to discuss to start preschool.

Really? asked Zabrasoma?

Really ! answered Tropical fish.

But you do not have age for school yet my dear Tropical fish?

Does not matter I will go to start this preschool and to start regular school like younger fish.

Ooo Good for you Congratulation said Zabrasoma.

Thank you very much replied Tropical fish. With those words Zabrasoma started its swimming faraway, while Tropical fish was continuing its way to its friend yellow tank fish. In its area around were living community of yellow white fishes or yellow fishes, or some others fishes that were close to his community.

When he saw some white in silver color body fishes with one elegance yellow frame decoration on their tale he did not get wonder but he thought with itself: Ooo my God of the Water ooo my honor POSEIDON, look how beautiful are all those fishes in our yellow - white family. Really they were so beautiful while their name called Silver fishes underwater.. those silver fishes were swimming in small crew. So after they gave greeting to Tropical fish they asked it why was alone?

Tropical fish said he was coming around while now was going to meet one friend. At that time Tropical fish asked Silver fishes were they were swimming like crew?

One of them with full energize answered so fast and with full smile. We are going to party of birthday of yellow Chichlid fish, while are coming all its cousins from yellow fishes from Chichlid communities.

Tropical fish, was explaining to them about its idea and proposed to them to swim together until to their party's place, while they agree. When they arrived over there they can't believe that in front of them will appear the most beautiful view, through blue clear color water was like painting yellow color of one big crew yellow so beautiful fishes.

Both Tropical fish and Silver fishes underwater made one strong uuuaaaaauuuuu with full wondering and happiness. Really is so beautiful view said Tropical fish, I did very good job that I swim with you here in this place. At that time all Yellow Chichlid fishes created one big circle around Tropical fish and Silver fishes underwater and they started to ask loudly so many different questions in same time. Tropical fish started to laugh and said this is the most happy and so beautiful fishes' crew that I have seen.

Tropical fish was seeing Yellow chichlids fishes and was whispering to itself: how beautyiful they are, how is created the nature to them green eyes and yellow color, I am proud that they have some connection with our yellow - white color families,really they are so beautiful. At that time he started to discuss with them about its idea for school, and they aproved its decision, they gave some alternatives, that if the center of preschool will be ine secure area of water by wilder, they will send and their small fishes to be over there.. But Tropical fish said to them that never it ihas heard that in its area of water to have any wild and that it never saw ony one. They got so wondering while so many of them said with one voice: Really you never saw any wild fish of water?!

Never answered Tropical fish. Topical fish did not finish its phrase of speech when heard one strong voice, that made it to turn its head. Tropical saw behind its body one big green white strips color fish its name was Perch Chichlid Hellow you Small Tropical fish, I am very happy to see you here, but iam getting shocked how you are alone here?

Hello said Tropical fish I came with those Silver fishes, but I am going to see my friend yellow Tank fish to discuss for pre - school, that I think to start.

Oooouuuu but do know your parents that you are here so faraway from your area?

Sorry, they do not know they will get scare if they know that, but I saw school fishes and after I met silvers fishes, that we came together here, so I decided to swimm more longer way and to see this Yellow beautiful Chichlid Fishes.

Okay so good, but you are young and you will come with me to tell something that is so important for your life Okay?

Okay I will come with you, while Tropical fish got scare, because it never saw this Perch Chichlid and knows nothing about it, but did not comment anything more. When they started to swimm all this huge crew created by Yellow chichlid fishes and Silver fishes underwater came and swimming behind it. When Tropical fish saw this action he got courages and was glad, so leader of this trip was Green white strips Perch Chichlid, and behind were Tropical fish, Yellow Chichlid fishes and Silver fishes underwater. After some minutes they entered in one area where the water was dark blue. All stopped. Perch Chichlid fish spoke loudly: Tropical fish come here. Tropical fish without voice went close to it. All others behind did not send any sound of their voice was full silence.

At that time Perch Chichlid ordered Tropical fish put your head more inside water.

Really Tropical fish got scare but has no choice and put its head more inside the water.

Look over there what are you seeing?

I see some fishes that never I saw in my life before they have different forms and are staying in distance with each other. Good you saw now, I am telling you what they are. At that time Perth Chichlid gave one whistle and all fishes with strange different forms for Tropical fish's eyes came close up to surface of dark blue water.

Perth Chichlid started to tell toTtropical fish, all are danger for you and all others, are wilder of the blue water, this is Seahorse fishwater, while this is Lionfish water, other is fish Exotic underwater that I do not understand why this beautiful fish is staying with those wild fishes, but is fact, so other is Tiger fish and the last is Scoprion fish so do you see them? Those are so danger for so many small like you and those behind you plus are some other creatures more danger and more big in their body that you never can image and they do not account how beautiful, young or good you are but they will kill you without thinking. Do you understand me now?

Yes I understand you.

Okay I never wish for you to see in your area those wilder biggest bodies because absolutely your life will be in danger for that reason your parents chose that beautiufl peaceful water place only for you to live in peace so the meaning is that you never will travel again with stranger in other areas without your parents or your close family.

At that time Perch Chichlid fish called Red Head chichlid and Blood Parrot Chichlid while was explainng to Tropical that those two fishes has some problems with that community and for that are staying in distance. When they came behind them came one very beautiful colorful crew fishes and one single beautiful fish too, different by them. Tropical fish was happy for new habitants that he was seeing plus it got lesson and news about those wilder that its parents never told to it. Tropical fish thought its parents loved it so much and they did not want to terrify their very lovely child with those scare figures for its life, for that they have spend all their treasure to establish their life in peaceful water place, At that time Tropical fish felt that it loves so much its parents and was happy that it will decide for starting preschool and more happy was for this big crew that were around it while were giving pleasure with their so much beauty, by their forms and colors

At that time Perch Chichlid called Red Head chichlid and Blood Parrot Chichlid while was explainng to Tropical that those two fishes has some problems with that community and for that are staying in distance. When they came behind them came one very beautiful colorful crew fishes and one single beautiful fish too, different by them. Tropical fish was happy for new habitants that he was seeing plus it got lesson and news about those wilder that its parents never told to it. Tropical fish thought its parents loved it so much and they did not want to terrify their very lovely child with those scare figures for its life, for that they have spent all their treasure to establish their life in peaceful water place, At that time

Durime P. Zherka

Tropical fish felt that it loves so much its parents and was happy that it will decide for starting preschool and more happy was for this big crew that were around it, while were giving pleasure with their so much beauty, by their forms and colors.

Chapter 2

Corals's fishes

When the Tropical fish decided to continue its way, after gave greeting and thanks to Perth fish for its advice, he saw very close to it two fishes that Perch fish called, like Read Head chichlid mixer color yellow with easy red, and Blood Parrot very yellow color both beautiful fishes. Tropical fish got wondering about their beauty and started to think that its community or its yellow family is so beautiful all around the water world. Both those fishes gave greeting to theTropical fish and they contiued slowly their swimming to go to Perth Green Fish. During its way Tropical fish was listening one whistle by one stranger fish that never has see in its egistence. It was swimming very quielty because did not want to break whistle with happines to this stranger fish

Tropical was swimming close to this stranger that really has one different form from classic form of other fishes, but so beautiful. When they were in front of each other the stranger fish gave one greeting and started to move its arm like segments like welcome to Tropical fish, at same time with one very interesting sound like fsssssssfsssssfsss,started to speak while tropical thought this fish is "Thuthus" that its mouth can not send out fast straight words. At that time Stranger fish understood its thought and said listen me I am not "thuthuq" but I am very happy and I want to make jokes while iam making my introducing to you about my name. all are calling me Single fish but some friends are calling me with nickname that they put to me like Fish Sword arm, while started to laugh with happiness. Tropical fish got wondering by its peacefull speech and nature that was in conlfict with its stranger and something scare apparance. But what about you rname?

Ooo my name is Tropical Fish. I am coming from other community but I have a lot fishes from my family here as I saw.

Ooo yes, you have a lot here because I know one very young small Tropical fish that looks like you, maybe is your brother? No sorry is not my brother I am single fish – Lazy boy in my family as are calling my friends and my aprents, but is my first cousin at that time Tropical fish started to laugh for his lazy nickanme so Single fish laughed too. Tropical fish said to Single fish may I ask you something? Yes of course?

What is origine of your family that you have different form from our community? Oooo I am explaining to you right now. My family have lived before faraway to one area that were so many rocks underwater, so our form is created in time by adopted of my ancestors to environment. For this I have this form of my arms and tales. Our family's form was in function of living to survive in wild environment., now I am going to meet one friend of mine that is living to one area with corals. Really we have a long time with two generation our family's friendship. Its parents were since youngest time in those wilder corals friends with my parents. Now I am continuing this friendship and family's tradition, while when I am happy or I have any very improtant news or beautiful story I want to tell to friend of mine to those corals, its name is Fish surgeon, really it is surgeon in its area, while is specilized for youngest fishes. Tropical fish was happy when it heard about this tradition and longer older friendship fo those two families. Tropical fish asked Single fish if he has any important news or beautiful story to tell to its friend that was rushing and was so happy. Yes I have one very important news to tell to it. Tropical fish do you want to come with me to see this corals place is not so far so anyway you have time for your trip, you will see so beautiful things over there that you never saw before. If is not very far away I am coming but I do not want to lose my time because I am going to see one friend of mine to discuss about starting school. So good for you to start school but come to see this strange relieve because when you will start the school you will have not so much time to come around because you will study. You are very right, so let's go together to see that.

At that time both fishes, Single fish and Tropical fish started their swimming straight to corals while they were discussing all the time and time by time Single Fish gave whistle with happiness. From the distance they saw that place that one colorful view was appeared in front of them while the water looked like was giving clarity to those colors of Corals. Tropical did not see before so beautiful corals and small stones multicolor, so it did not hold itself but gave one very strong cheering with one longer uuuuuaaaaaaa.

Those corals really were so beautiful and so many others youngest and smaller multicolor fishes but most of them yellow were coming around. Tropical fish asked single fish: Why are all those youngest fishes here?

I told you that fish surgeon is doing very good job to those. some of them has birth defect some has got damaged during moving and this Fish surgeon is doing art improving in their body because it said that everything is easy when fish are youngest than adult to work with their bonds or whatever, so they came here to put appointment with this fish surgeon or some of them have day today for their surgery as saying in new time now plastic surgery. Their parent are taking care so much for their youngest and they are sending here. When they went more close they saw so many different colors fishes like pink, blue mixer blue and yellow and most of them yellow. Tropical fish thought fishes in this area, are more careful for their appearance I am proud for those families and this community, look how much are taking care parents - fishes for their kids –fishes., while it felt that its body was full with happiness

While Tropical fish was thinking about this matter heard single fish that was greeting loudly one blue in purple color fish that they called it Fish surgeon. This fish surgeon was coming fast full glad for its guests. Fish surgeon while gave greeting to both of them and after Single fish introduced with Tropical fish, asked what is problem of this happiness for Single fish.? Single fish or Sword arm fish has called its friends in its nickname answered with big smile: I have one good news. In our area came one project to do one experiment for rising new specie of fishes for our water – area but those specie are coming from salt water, so the community decided for me to lead this project and experiment.

Oooo I am very happy for that dear single fish. Congratulation I know that you are very smart and persistent fish, I trust you will have huge success.

Thank you very much dear Fish surgeon but I have one problem or more exact our community has one problem that we need your helping?

What is this? We need you help to meet Fish emperor to give permission for us to start our experiment to our area, because we need this permission to possess one part of underwater. Dear Single fish I am giving to you in presence of our guests Tropical fish that I will help you and I hope this project will have performance. Single fish started with humor its fsssfsssss that was sign of happiness and said Thank you so much. While Fish surgeon knew very well single fish said: started you now with your glad fsssfsss and laughed, after invited both of them Single fish and Tropical fish to swim together and to see the beauty of colorful of this area. All were happy they started together to swim and discuss careless but will happiness for different problems.

Chapter 3

Pink Fishes' community

During their conversation Tropical fish and single fish make with sign each other when one very beautiful yellow mixer with white color youngest fish was swimming close to them careless, while they never interrupted the conversation of Fish surgeon that did not give attention to this young beautiful yellow Chichlid fish, he was speaking for project and science. Tropical fish said with low voice to Single fish, this youngest beautiful fish is from my family, they are living all around. When the Fish Surgeon sent those two fishes in border of its water area, it said to them: I will decide one day to meet both of you because I liked Tropical fish for its patience about my conversation and I want to tell you one place water that will be with big interest for your new project single fish. They saw each other and they said okay we will see you after one week in same place to go together to this new water area. After that Single fish and Tropical fish continued their trip together very enthusiast about this meeting with this very important fish. Tropical fish started to explain to Single fish about its situation. It Said that is very young to understand all those situation that are for adult fishes, but I like adventurer and I am very curious to see and to know. Single fish said is a long time you to arrive in this phase but to see is good for you because you are child but I need to meet your parents too, I think to take them in this meeting with fish surgeon next week. Child Tropical fish approved without sending out any voice. While the Single fish or Sword arm fish came to its area he hugged the sweet Tropical fish and said it will go to its area to meet its parents and next week all together will go to Fish surgeon. Tropical fish was very happy about this occasion for its parents and smiled with happiness. It continued its way to go to other friend to tell about this preschool. After some times of swimming alone it saw in distance its friend that gave sign and was lifting up the water that looked that was very happy. Tropical yellow fish was thinking about that while in front of it the blue color of water suddenly changed in pink color looked like one big tank was thrown all around the pink painting liquid on water while the sun's rays were entering in specter through water. In front of this very beautiful pink water was coming one magnificence fish that all called it Tropical Fish Hook.

Really this fish was so beautiful his pink color was alternating in yellow color in middle body and after the yellow color was alternating in white color at bottom of the body. Tropical Fish smiled and said really its friend is very beautiful, behind it were coming one very beautiful bold pink young couple that during their swimming they were kissing each other time by time and they did not care who was seeing them too.

Others were calling them Pink fresh water fishes, they looked lovely couple and they were taking care for Tropical Fish Hook because they were friends of its family, so they do not want to leave it alone all around. When Tropical Pink fish hook came close and gave greeting to Tropical fish, it explained to it that this couple are very lovely friends of its family so they do not want to leave alone it. Tropical fish understood and agreed with that situation. At that moment when the lovely beautiful youngest couple were coming in front of Tropical fish suddenly one strong white wave splashes their face and body and thrown in front of them one very beautiful Pink Fish star that looked shocked and was moving very scare its five arms without direction looked that wanted help.

At that moment Pink fish star fell with force by waves to some small stone while sent one scare lost sound that terrified the youngest couple, they yelled and went to help this little pink very beautiful small pink fish star that looked without breath. They got together two its arm and moved it to see if was alive. The Water god has created the most beautiful pink fish star that through its body and five arms above pink color were spreading in uniform some yellow white small sign like spoiled the skin. Tropical fish made one strong Ooo my God how beautiful is this small Fish star. The Pink fresh water couple screamed from happiness when they saw that small pink fish star was alive, all were happy for its survive from hitting by white waves to stones underwater. Pink couple fish said with one voice: Only Water God, the honor Poseidon saved our very beautiful pink fish star.

When the pink fish star started to move its five arms all were quiet and they started to discuss together. Tropical said with smile to Tropical pink fish hook that they together will go to start preschool. Hook and couple fresh water opened their eyes and moved their tales by wondering, but they asked, when,? And where? Tropical fish said I have decided both of us to start our preschool and with advance program to move in more high level of school, because I got inspiration by one crew fishes close to my community where they were coming out of their school and they looked so happy. Okay I agree said Hook fish so approved and Couple pink fresh water fishes. At that moment their quiet conversation interrupted by one loudly noise by one huge very beautiful pink crew fishes, that were swimming straight to them. Their pink crew looked like one flux pink light through blue color of water that wanted possession of this underwater area, One magnificence view was created. There were Funny Neon Pink Fishes Crew. One of them started to whistle with one very characteristic sound like zs zs zs zs zs after started to speak with happiness: sorry for not properly behavior but as curios fish I am I was listening in hidden way your conversation and I called all my family and cousins to come here like crew and to say to you that we want to come in preschool together with you anytime wherever to be this school. Pink fresh water fish were happy so was Tropical fish Hook and other Tropical single fish. All agreed with their very good decision and Tropical fish promised to them that will give news to them very soon.

Chapter 4

Sharks

Tropical fish left behind all this beautiful pink youngest crew fishes, also he left with sweet greeting and was continuing its swimming straight to its area to meet its parents. When Tropical fish arrived over there all its neighbors and friends were in alert situation and full frustration because was a long time that they did not find Tropical Fish. Tropical fish felt desperate itself for this sadness that created to its parents friends and neighbors, really it was swimming faraway and no one has news about its trip. When those saw it they all screamed with happiness and said came Tropical fish came, Tropical fish came and we are happy that it came in good condition, Its parents stopped their swimming while their yellow body's color close head lost its shining because of their tears that were coming down and mixer with water blue. Really little Tropical fish felt itself so desperate. But when he hugged its parents with its transparent arms everything came neutral. After that when its saw all others fishes created one circle around it, with so many questions in same time, Tropical fish started to explain about its wonderful trip and what it saw during this trip. He started is story:

I was swimming today alone suddenly I saw one blue gray with yellow tale, crew fishes that were coming out of their school so I decided to go to meet friend of mine Tropical pink fish but during my swimming I met and saw some new friends fishes and was my big pleasure when I saw one very beautiful community yellow fishes Chichlid family, my dear parents they were from our yellow big family. After I met Perth Chichlid, I heard about one experiment of vaccination in that area also I met Stranger Sword Arms fish, or Single fish that has one big project, I got introduced with Fish surgeon and so many others also with my friend pink fish. I saw the most beautiful pink community youngest fishes that wanted to continue preschool like me, now I am here but I have one good news for my parents: Fish surgeon wants to meet you and to talk with you and Single fish or Sword Arms fish about this huge project for creating one new community of fishes that are coming from salt water, they want to adopt in our water blue. What do you think my dear parents?...while I apologize for this scare situation that I created for you today I know you do not deserve that. Okay Dear Tropical son fish they spoke in same time: We will think and discuss for this problem before they to give signal or news to us. After all were happy and started to discuss careless. The time was going fast when one beautiful sunny day came to this area with speed Single fish or Sword arms fish its nickname. All youngest got wondering by its form and some of them got scare and why Tropical fish has told to them. Single fish met Tropical fish and its parents. They discussed so long after they decided to go together to meet Fish surgeon also two friends of them proposed to go with them so those "Five" fishes started to swim to the area of Fish surgeon. During this trip they saw all those communities that Tropical Fish saw before. They fixed those beautiful view in their celluloid of their eyes and after they stamped those pictures in their body while they were helping each other, so their yellow color of body came with different view that looked like map or like artistic tattoo. When they met fish surgeon he started to laugh because he saw those pictures and said maybe is first time for you in this long trip.

Yes they answered is first time that we are swimming so faraway. But at that time one loudly noise came from distance. They turned their head when they saw some big figure open gray and dark grey color. All of them were speech less. Fish surgeon said calmly: They are wilder Sharks, but I do not know why they are so mad today stay quiet do not move and speak also do not give reaction whatever they to say or to do action because I am here they need me, so they can't do any danger situation now, but they are so danger. Three sharks came straight to them but in normal distance they stopped their swimming and were coming up and down in water while started to yell:

Who are those that wants to do this project in our area blue water? We got news by others, What they think are? We are power here no one can do anything in our area.

Fish surgeon answered to them quietly: First of all they are coming from one peaceful area and they do not know you and your power too. Second they came to discuss with me but the project they will do to another area water blue, and is doing single fish or sword arms fish but they are intellectuals and want to help it.

At that time Three shark madly started to splash water up and they sent water in direction of this group, while they started to yell again: Does not matter if they are not doing huge project here, wherever they will do this project they need to put our name we must to participate in this project and to have our profit, because we are power in water blue for while we never will allow them to do this project without us but will sabotage and undermining so their project never will see light of the sun through water blue. Single fish opened its sword arms and said: Who are you that wants to take my intellectual project, you are wild but not intellectual so I can't accept your proposition:

But they started to yell and said again we are power here, we possess all this water blue also we can continue school like you in modern way in short time so what?!...we will be same like you intellectual but more better than you because we have power, so power plus intellectual we will be more stronger than you, after that we will prepare one strong strategy to lead all this area with blue water in modern way with some modification not like now, so life will change for all fishes, but only us will be main leader.

Single fish or sword arms fish said strict: Wilder are wilder I do not believe in their idea of modification of the way or in their changing, I can't believe my peaceful project to wilder, my project is for life or prolife.

One wilder shark that was more young and more capricious said: Okay you told us that you have high level of education and are intellectual, what kind of intellectual are you that do not accept progress of others, changing or modification?! In our planet so many things have changed in centuries wilder came more soft and so for us, so or you are intellectual or you are not intellectual, you are not accepting the theory of Biologic science honor Darvin that spoke so much for us and for flora on planet, that everyone kind of specie that is not adopted with environment has destination to disappear, so and for us we will adopt with this new life and will exist so what now? We need to be part of this project.

All were speech less, but Sword arms fish spoke with persistence way: I do not believe in wilder and in their changing for good that is it.. at that time they said with one voice: We will call all our wilder families as you are naming us and will protest against you.

Do what you want said Sword Arms fish, because we will do this huge important project so far away from your area of blue water, so your power over there will not have effect. They made with irony pz pz pz pz while where showing their big teeth and splashing water up after they turned back and started to swim. Parents of Tropical fish said, they are danger so is one big issue:

Sword arms fish said: It is not any big issue because of them, they are danger but I am not scare by them, they want profit because they want to eat in easy way, so I will continue my huge science project with all others fishes with courage and with you too, so do not think for them because all the time they want problems only to show their supremacy. Do not care for them, but now let's go to see that area of this blue water we should do this project. Fish surgeon started to laugh and said Single fish is very right we will do this new project for creating new community and rising new species of beautiful fishes that are coming from water without salt.

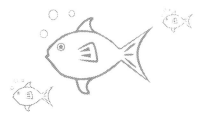

Chapter 5

Preparing of Meeting

Fish surgeon with Single fish or Sword arms fish thought very serious about this project while they did progress about preparing situation for big meeting with different fishes communities. Time by time they did back and forth trip to Tropical fish's area with a lot discussing with its parents. Time by time they were doing meeting with Perth Chichlid that was busy with its science's center of vaccination. After a lot preparing and discussing they decided day of meeting. First of all they did one meeting only with the representative of those different communities, one problem was about claiming by sharks to attend that meeting, while they discussion why whales to attend and they not, so their attending was pending by Fish surgeon and all others in this board? Every one of representative gave their idea about this project, but one big issue and a lot discussing was about fund for this project, at last they decided to work with donation that everyone fish young or adult to give a little bit amount of their food while they will sell one of their scales. They decided that based in beauty and selling of scales will be the amount of donation. So one scale that every fish will sell will not damage them while gathering of those profit by all fishes will ensure amount for this project. They decided to take like guardians for this meeting some blue whales that will control with their jumping out and in water all area.

The day came of meeting while the sun was shining all around and its rays were going inside the blue water and were making light place of meeting. The fishes were coming in crew. This meeting was open without limit. They will listen speech by vibration of voice through water in big distance. The sun will help with its rays to give voice the speed of sun all around that was new technology for all fishes. The fishes from all communities were coming to see and this new technology in their world underwater.

All around the white waves of blue water were doing advertise in crew while were dancing with happiness and were whispering their advertise all around for this meeting. Some white pelicans in crew with some white and pink two colors pelicans became jealous for this unusual moving of waves and they flew above the blue water while they tried to put their head inside water to see what was happening over there. Sometimes they were using their arms to splash water and to see inside more clear, really they wondered by this moving and signing of white waves but they did not understood, they saw so many fishes that were swimming in big crowd in same direction but they got confuse what was all of this.

So they decided to speak with their friends alligators and why they were wild also they thought to discuss with amphibes, like their friends frogs. They thought frogs are more peaceful and no one will control them so they will investigate, while alligators are not welcome by most of water habitants. So they flew to one place that water was creating channel and was connecting with ocean blue water. They went over there and found some frogs that were taking sun's rays on their body that half was under water half was out on ground without their green clothes. Pelicans thought: what is this strange day that frogs are nudo only in their white inside skin, on this channel but they did one voice and frogs moved inside water so they dressed their green clothes and ca

They told to Frogs, about strange moving of white waves and different fishes in unusual way, so they wanted their help to make investigation what is going on over there and to tell them. One younger alligator came close to them listened their conversation and said why they did not said to him to do that investigation why they did not trust to it that action. Pelicans did not want to offend this younger alligator for its wilder nature to others but they said that frogs are so small and no one can give attention to them so they will do quiet investigation what is going on over there. After that alligator accepted in silence this version of their decision and started to move away without say greetings, so they understood that it was like break heart. Frogs were happy that it left and said that: We should put distance with all alligators anyway youngest that looked innocent and adult because they all are same wilder inside.

So after that conversation they decided that some frogs to go inside and to do investigation inside water to that place where was big moving of white waves and different fishes in crew. Frogs got prepared with emergency while started their swimming deep in underwater. They saw one big moving by different fishes that were discussing with full energy about one huge project and one new technology that hope to see in this event. They saw so many new form fishes and colorful that they never saw before and they got wondering. Frogs never believed that were existing underwater or more exact in the world water so beautiful fishes multicolor with different forms. Frogs got shocked while they said: Oooo our Honor god of Water what beauty you have created in this water world. They coped in their eyes what waves were writing with their white color liquid inside water and wanted to give this description to Pelicans for translating but as they understood, was one big event of fishes with big importance. They did this observation but they were wondering from this beauty and they decided not to go back fast, but they wanted to see this beautiful parade of those multicolor fishes. Frogs thought: Why we do not make some events in our community but only we are doing kuak kuak kuak for nothing all the time in spring time, we must to think to make big things in our community look fishes what they are doing, they are not more smart than us, plus other science's habitants said that Fishes' brain has not so many wrinkles that is making it selves not so smart, so we need to study and to move to do something for ourselves. We should do something for our big community of Frogs all around. We know that some habitants in ground like our song in spring time but we need more attention and more respect by others so we need absolutely to do something to do for all of us Amphibes.

Chapter 6

Long swimming to event's place

That day of the event of free fishes for big project was unbelievable how many different communities of fishes came and their diversity of colorful was giving to clear blue water one view that was creating image that rainbow came illegal inside the water to give flus colorful light to fishes. During this pelegrinazh of fishes under water two stars fishes blue and red to orange color, were discussing strongly about this project, while behind them were coming in line yellow Chichlid fishes, with so many different groups that have some other decoration around yellow body but all of them were Chichlid yellow family. Close with this group were swimming the most beautiful Blue fishes group they were speaking loudly that they were claiming to be in this project.

Behind this Blue fishes group was one crew that they have created one circle form around one fish that looked was Fish surgeon. They were so enthusiast while said to Fish surgeon that they should do this project with ever prize fish with sword arms was speaking loudly and looked that wanted attention of all others fishes. The Fun Pink Fish was coming with this circle group but behind it was one huge so beautiful pink fishes' crew, that really were so silent, that never were loudly like yellow Chichlid Yellow fishes, and Blue fishes. Pink fishes were swimming quietly and some of them were speaking like were whispering, really were calm fishes, strangely all others were seeing them and get wondering how quiet they were, while all know that those pink fishes were so beautiful compare with all others colorful fishes.

After beautiful Pink fishes were coming one crew of Green fishes that strangely were more big bodies than all others. Those Green fishes looked that got shocked by all of this view between all different others fishes.

Some fishes that were swimming in small crew but in different way, they were swimming in form of circle and their head were touching each other, looked them were speaking in hidden way like secret way.

There bodies were two colors orange and white so beautiful.

Really looked that those Orange white fishes did not care for others fishes but only for them conversation during this trip – swimming to the place of this big event. Not in big distance by those orange white fishes were coming loudly and so much careless one very beautiful orange crew small fishes.

For their beauty and blue water got wondering and satisfaction while gave reaction with moving itself strongly up and down, at same time orange small fishes understood that reaction and started to smile. Behind them was swimming so slowly one Sea turtle, all knows that was coming from faraway.

Turtle has done one very longer trip with its slowly swimming to attend this big event about this huge project for rising in that area one new community of fishes that were coming from salt water in that area.

No one understood why some small grey - green color fishes were covering this small turtle while they were swimming above its body and were covering turtle in strongly way also one big crew of this community was swimming behind.

No one was understanding this situation of this turtle, while all other groups of fishes started to talk with each other during swimming and were watching with their eyes' corner this quiet turtle. Suddenly one terrible noise came behind Sea Turtle from inside the blue water and one strong moving of water in surface came like from one strong force that was lifting water up in wild way. All groups of fishes understood that something so bad was coming to happen. They started to ask each other what is this strong terrible noise, because so many of those beautiful small colorful fishes never saw wild life and wild fishes in their life, they were coming from peaceful place in community with very peaceful fishes. When they were speaking with big wondering they did not finished their speech when close to them came one wild, madness sharks' crew, that were yelling and screaming all around while were giving one big noise that terrified all smallest fishes. Their leader of each group spoke with each other and said now we understood why Sea Turtle is getting cover all around by those great fishes, but now we must to establish this situation and to make calm our youngest and smallest colorful fishes that are terrified by those sharks. They called to each group their small fishes while were ordering and advising to stay together no one to be separated by others and to leave one space free of water that those mad sharks to swim freely and not to create conflict with anyone small beautiful fish, because they are so much danger, plus now that are so madness their danger is ten times more strong. All colorful fishes approved without speaking and started to create one line while they were placing it selves in long line where each row will have four fishes, and the free space was creating too. Grey Sharks came with big noise while their madness was going up. They got news about this event and they were so mad, why fish surgeon did not tell them, why Fish surgeon did not invite them. Fishes' leaders thought that was not their fault that fish surgeon did not call them and to invite to this very important event. So they decided to stay quiet and never to involve in this conflict between powers, because anyway they will be losers. So they showed skeptical behavior to mad sharks and their unreasonable noise, moving and their noise too. They opened the free space for those sharks while they never turned their head to those wilder but let' them to swim in that free space. One silent situation was creating all around looked that no one fish was existing in that water – area. Sharks were swimming in that place full with loudly voice and they did not care for those small fishes, they were swimming and they demonstrated like all the water was their property, in their own, but was not true because sharks can't own all that blue water. One small beautiful orange fish said:

What they think they are? But its leader made with sign with its transparent arm like do not speak and Orange small fish did not speak anymore and why he understood that something was wrong. At that time its leader said to small orange fish, you are very young, so never you can't understand that situation does not matter that you are right, but in our water place's life are happening so many thing that we can't understand or can accept but is not in our hand so we need to adopt with situation. Suddenly and strangely beautiful small orange fish said in strict way that shocked its leader: I do not understand that phenomenon to adopt with bad situation what is wrong is wrong and must to be right so I understand life in our water world and specific in our area. Thanks to Water- God that we do not have wilder in our water area, but now I am with you and all others and only in that trip I will adopt myself with strange situation but not more longer but I will demand and fight for right things in our water - place's life.

Good dear orange small fish but now we need to be silent and to escape this very strange and very danger situation by those wilder for all of us that are going in this very important event. I am very happy to attend this event finished its speech the beautiful small orange fish. Yes you are right approved its leader while said, I am happy too to attend in this very important event for this huge project.

Chapter 7

Zabrasoma's speech

Yellow Tang Community Fishes

Through the most beautiful colorful audience created under blue roof water by those different fishes 'communities the Fish Surgeon went to his stage position to start the historic big science's Convention of Free Fishes. At that time all yellow fishes were in front of others, behind them in line were green fishes, after green fishes were in line orange fishes, were continuing this intercalary colorful communities by pink fishes, behind them were blue fishes, blue and white fishes at the end were red fishes and some others mixer colorful fishes. Strangely when the sharks saw with big wondering that all those fishes, were establishing it selves while looked that one fish was putting all of those fishes by their color combination Sharks did not speak they stopped their noise, and why they faced skepticism by all fishes, they saw each other and they gave signals they will stay in this Convention while they will be aside all around. So when all fishes were established, all sharks in distance without noise established it selves like fence all around.

Now I will call name of some fishes that will be representative for all communities here to be in stage aside me. Fish Surgeon started it's speech with one sweet welcome for all fishes that were attending this convention while he continued with words:

Please come and enjoy me in this honor stage honor Yellow Tang "Zabrasoma Flavescens at same time we will operate in this way everyone leader here will introduce itself and all its community. At that time the beautiful yellow scaleless Fish came in front of page with sweet smile while all other fishes in audience with strong applauses their arms while the blue water was coming up with sparkles. Zabrasoma Flavescens started it's speech. I am Zabrasoma, my family is in Yellow Tang fishes species, that are swimming in salt water.

Family's name is Acanthuridae. Our family's members are so beautiful and are so liked by others habitants of this planet, that wants us to use like decoration in their some strange assets that they named those Aquarium. Real we are ocean swimming fishes and we have this problem for isolating us in aquarium like prisoner, this is not fair because for our beauty we do not deserve this prize but we need freedom in our life. So many our youngest beautiful Yellow Tang - Fishes are isolated in those assets by habitants of Earth planet, to give them pleasure and to make beauty their living place as they are naming home, or business working that they are naming those business's office.

So we need their freedom more exact we need for our beautiful yellow youngest fishes freedom.

This is one of some matters that I need to discuss in this Science's Convention. Habitants are naming us like popular fishes of Aquairum. I am proud that my family is yellow tang in "Surgeonfish" family like our leader "Professor – Doctor" of this Convention. Our family's adult fish can grow to 20 centimeters, (7, 9 inch) in length and 1-2 centimeters(0,39 -0.79) in thickness. Adult males tend to be more larges than females for this reason those male fishes gave priorities to their activities and are humiliating females fishes, that is not fair. This is another matter that I need to discuss in this science's convention of fishes. Our family Yellow Tang Fishes have bright yellow in color. At night the yellow coloring change while is developing in the middle horizontal white band, but during the middle day they are sparkling their yellow bright color.

Our yellow tang fishes in wild nature water are feeding it selves with algae, also some others marine plant material too, but in future we will have problem about feeding ourselves because number of fishes is growing up food are destroying by so many factors and we can't follow in specific way our diet nutrition this is another matter that I need to discuss in this Science's Convention. Some expert are thinking for us to feed with animals meat that are producing complex amino acid and nutrients that only ocean animals can provide, but this is future science. I want to make remarkable about our very good job that we are doing in our community like community service, we like family have specialization in "Cleaner Service" that in wild nature in water we are cleaning algae that are growing to shells of the Marine Turtles.

As all know we are fishes of salt water, but some science habitants of Earth planet are creating some good condition for us about low level of nitrate that is damaging us, is creating high level of nutrition and high level frozen of protein, so we need to study about those, elements in our water kingdom and to transfers our Yellow Tang Fishes community in water that has low nitrate but so much protein by animals. So needed science in our life for wellbeeing. Our longer life is until to 30 years in wild nature water but our youngest has in aquarium of habitants of Earth planet in small Aquarium with 50 gallons show tank size life is 2-5 years, in average Aquarium with 75 gallon show tank size life is 5-10 years and in larger Aquarium with 100 gallon show tank size, life is longer for our fishes until to 20 years. Those facts are tragic for us that our fishes have short life because of their isolation and their full lonely desperate and why their food is prepared by science. This is another important matter that I need to discuss, because habitants of Earth planet are invading us, our new generation. We need to find one solution to protect our youngest beautiful yellow fishes. Some of our youngest yellow fishes are "killed off" in first months of care from hobbyist mistakes, inappropriate tank mates starvation that is very bad phenomenon and we must to stop that action careless.

Our yellow tang fishes are with smooth skin scale less, semi aggressive, they are swimming in pair or group, they are taking poison by high level of nitrate in water so they need one area water with measure level 30 parts per million (PPM) in whatever ecosystem the tang in housed in. All Yellow Tang Fishes are quiet susceptible to "Cryptocaryon Irritans" or "Marine Ich" a parasite that named "Freshwater Ich" also and some others common saltwater diseases. That really we need to protect our community with some new vaccines. About "Marine Ich" we have avoided that, because we have provided with science in our area dropped of temperature. Some our famous yellow fishes are Cardinal fish, large Clownfish, Lionfish, Eels, they for their famous name and exceptional nature have privilege while are more safe in our water's area also and in habitants' Aquarium, but we need to create safe conditions for all our fishes without selection.

Really we or all of us Yellow Tang Fishes are very peaceful inhabitants that is an important part of our ecosystem. But we are abused by some habitants of Earth planet that are using us for their big profit. We will fight to protect ourselves and never anymore those habitants of Earth planet to abuse our beautiful youngest members tang fishes. We hope our life after this Science's Convention will come full with happiness peaceful and bright too. Thank you very much Dr. Fish Surgeon for giving me this bright opportunity to speak in this Science' Convention also thank you very much all of you for listening me.

At that time all fishes gave one strong applauses while the water was coming up like waterfalls was full happiness by all and so many expression by different fishes for that high level and very beautiful eloquent speech by Zabrasoma Flavescens.

Chapter 8

Orange Fish

The Surgeon fish came again to the stage and called the name of Orange fish: Please come to honor us at that stage Orange Roughy Fish!

From the big crew of orange fishes came this beautiful fish. It was appeared with pompositet with its actually a bright, brick – red to orange color, however, the orange roughy fades to a yellowish orange after death. It introduced itself. I am Orange Roughy Fish from Hopostethus Atlanticus family. Really we are deep sea fishes and we are belong to Slimhead family with our science name Trachichthyidae.

We are living in water with temperature 3- 9 grade Celcius or 37- to 48 grade F. in deep water to 180 to 1800 meters (590 to 5910 Ft water of Pacific Ocean, we came here in that area for exploration and we want to attend this Science's Convention. Really our family is opened all around planet in different ocean like East Atlantic Ocean, Indian Ocean, East pacific to Chile, South Africa's ocean until to New Zeland and Australia.

We do not have goal for expansion but our goal is only to study water and to find the more good ecologic for our family's members while others are underestimating us. My family members are so beautiful for their color and appearance so many loves us only for our good looking. But really I have one problem that I want to discuss here, because as you see our eyes are so big we can see so many things and why our mouth is more big than others fishes, but really we are speech less.

Durime P. Zherka

Some of us are small and some are big really our measure are 75cm (30 in) but average commercial size is 35 – 45 cm (14- 18 inn)with lateral line in uninterrupted line with 28-32 scales with weight 7 kg or 17lb, while others are identifying us as underused for less gastronomically- appealing "slime – head" so need more marketing for our beauty and our service in this planet and to be more important in their Fisheries program. This is our request in this convention. Habitants of planet are notable us only for our lifespan up to 149 years or more specific between 125 to 156 years.

Maybe is one reason about skepticism of others for us because we are growing so slowly for 20 years we are going only 30 cm or maybe of our characteristic about reproducing, that our fish female are producing eggs compare with others fishes, 22.000 eggs for kg of body weight that is less than 10 % of average of other species of fish that is telling that fecundity is low for ur fishes and this process of between fertilization and hatching is period time 10 to 20 days. We have another request about this process to help our fish female for more producing eggs that will make addition of our heritage, and this is science problem that needed to discuss in this science convention of Free Fishes.

Because we are here in this Convention I think another science problem for my family is about our length of adult age that is going in range between 23 – 40 years, that is limiting population growth - recovery, because each new generation takes so long to start spawning. So I am pleasing this honor Science - convention to help us with its science research program in future to help us about making more active and more short this phase by youngest baby and new generation so the growth to be in high level and recovery to happen more fast.

At that time one strong sparkle came from water up and some noise of fishes that screamed with happiness you are very right you need help after that was one strong applauses.

At that time Surgeon Fish said this was one great information that we are interested in our science about those problems thank you very much, Orange Roughy Fish.

At that time Orange Fish gave greeting with smile to colorful audience under blue water with its words:

Thank you very much for honoring me in this convention and this stage and for your listening too.

Chapter 9

Pink Fish's speech

After some noises while Surgeon Fish was preparing to call other name in audience one Orange fish was speaking with high voice while its voice was coming like whistle through water. I want to tell to this audience of fishes that we are Orange Fishes too, from famous family Symphysodon, that we have three big species of Chichlid fresh water fishes, why we do not have right to represent ourselves in this Convention or maybe that we are living and swimming in Amazon River and not in Great Ocean?!. But one small Blue fish replied to this Orange fish, of course and this is one reason that you are living in Amazon river and not Ocean, plus Amazon river never can compare with great Ocean because he is going through the land like it is sneaking not straight, right?!, plus has some damage from Amazon river in its way in all areas close to it.

Who are you that are speaking for Amazon river in that way?.., Do you know that are so many ships that are going through and transporting so much materials and people too in this river? Also Amazon river is making fresh air all around area where it is going with its water!

Yes, it is doing fresh water but like sneaking form that I do not like answered blue fish, plus I do not like its pride to compare itself with great Ocean, if it wants to be close with Ocean, Amazon river needs to make so much job for all of you under its water and for all habitants of this planet that they are complaining for Amazon's river devil mind, that wants only to possess everything and if, it can't achieve that goal Amazon river is doing flood like revenge that I do not like that, made remarkable blue fish.

It is good for you to think before you to speak for our dear Amazon river said Orange fish, because it is so lovely with us but as has Great Ocean good and bad or soft and wild fishes or water's animals so it is and for Amazon river finished its speech Orange fish while Blue fish that looks that has strong head but was pride for its knowledge and speech said:

Listen me you never ever will convince my mind about Amazon River, because I know that has so good points this river but has and some strong bad points about habitants, of planet, animals and ecology so needed more hard job by this very longer and stronger river but do not forget its traveling is like sneaking this is point of my thoughts, plus I am saying to you dear Orange fish continued Blue fish with irony, or more exact I need your answer. It is Ocean that is felling or sending water to Amazon river or it is Amazon river that is throwing its water to Ocean?.. heee tell me, so who needs more each other tell me?

Orange fish saw with skepticism but did not answer because it understand that Blue fish was persistent and was right, while at the moment all audience gave one strong laughing.

At that time the Orange fish turn its body and wanted to go to its community more faraway than stage while spoke with irony with half voice:

Amazon river is great, it is big, has so much water and it needs to send its water to Great Ocean.

What did you say, I heard your half voice said Blue fish, but I am telling you something, it is true that Amazon River has huge gallons or billions gallons or tons of liquid that named water like our water but really I never saw its water but this river is ignoring all others rivers in that area, so it is thinking that only its water is value, and never is helping other rivers with its water or never is appreciating those rivers for their job that they are doing in that area too. Definitely its selfish nature is punishing your community not to speak in this Science's Convention, because your river and your community does not work with "Science" only is making "Noise" with its water, so go to work with your community and your dear Amazon river too, when you will achieve any always I am saying "Any" goal come here to speak for your community and your dear Amazon river.

All audience laughed and said very right go to work and give this strong message to your dear big Amazon River. Orange fish was swimming quiet to its place and did not speak.

At that time Surgeon fish with smile called other name of Pink fishes' community. Now let's come in stage to represent its community honor Pink Clownfish.

Pink Clownfish was swimming quiet straight to stage while its appeared its pink color overall of its body with white line down the back and a white head bar. Pink clown fish started its speech. I am from Amphiprion Periderajon Family, all my family members are so beautiful like Anemone fish for that reason all others habitants in planet wants to use us like decoration for their strange aquarium in their home or their office as they are naming their living's assets.

We are like anemonefish so much sensitive with length until to 4 inches. We are soft and we are swimming so quiet without noise. I specific do react so strong negatively to poor water. I and all my family and the most of my community's members are claiming for quality water. So this is one point of my discussion in this Science Convention about creating water with some nutrition for my community while PH we need to be 8.4 to 9 that is another request of this Science's Convention. Generally we live in harmony with group in 100 feet deep of water. With temperature of 74 to 82 grade Celcius. Another characteristic of my community pink fishes is that females are big with their body than male fish. I personally tend to be dominated by my relatives. We all are less aggressive but other species close with my community that are entering in pink family fish are more aggressive than us. We are so beautiful with our color and our body we really have different size of body of fishes in our community, but our beautiful appearance has awaked big jealousy to others fishes that is damaging in some aspect our beautiful and quiet life. This is one very important reason that I came here with some members of my family to do request to this Science's Convention to give to us some information for others area with good quality water and with PH, that is optimal for our body because we want to move from area that we are living now. We do not like to fight with others but we love to live quiet life in harmony with our group. Others fishes are so jealous for us while are creating one big problem because they want to destroy our beautiful and loyal group of Pink fishes, so I want to save my community and my family, for that reason I need your help. We have created one board "United Pink Fishes" that leaders are our beautiful and lovely Pink Kissing Fish and Pink Betta Fish. They are working hard to keep unite and strong our pink community fishes. At last Pink clown Fish finished its speech while gave its greeting and its thanks to all.

Thank you very much for patience for listening me !

All other fishes gave onestrong applause while said to it: Keep safe and strong your family and your community.

At that time Surgeon fish said with high voice:

Good job is done by Pink fishes, I support their activity for United of their community and I wish and you will support it. Our Science's Convention will work with special program to study and to find for this very beautiful community one area with optimal PH in water and with good quality, while I wish for them good life and longer term harmony and love with each other.

All others fishes did one strong applause with their arms and gave one strong cheering. All this colorful audience under blue water was energize by this full optimism and happiness.

Chapter 10

Juvenile Northern Pike Esox Lucius's Speech!

Surgeon came again to stage and called **Juvenile Northern Pike Esox Lucius** from Green Fishes community. At that time one very elegant green fish that its thought about its impact for its beauty to audience was swimming so fast to stage. Its elegant thinner body with some green open to white color stripes that were like circle through its body, created buzz to this big audience by fishes. It started with full energy while it did not wait to leave stage surgeon fish that this started to laugh, with Juvenile Northern Pike Esox Lucius. It said my name is Juvenile Northern Esox Lucius. But some friend of mine are calling me in short name Juve – Lucius, some others are calling me Esox Lucius and some only Lucius.

I am here to represent my very beautiful; Green Fishes community, that for coincidence the Earth's habitants are leaving us because our color is in common with green color space that they love to create in big space as they are saying in different encyclopedia, so we are beautiful but we are very lovely and same time we are loved by water and by habitant on Earth. In our community we have so many different forms of green fishes also like others yellow blue, red fishes and we have green star fishes too. All know around for our beauty and our combination more about green and blue but today really I want to speak in this Convention about one specific famous green fish that it s name is: Green sunfish orscience name is" Lepomis Cyanellus". This green fish like all of us in this community is swimming ad living in freshwater. It is coming from sunfish family

"Centrachidae" for its beauty habitants on Earth are using like decoration in Aquarium. Our green Sunfish has one very specific point that has so much value for this Science's Convention.

What is this value point?..asked strongly one Blue fish from audience without taking permission or saying please for this interruption but never said apologize too. All other fishes understood that this Blue fish suddenly became nervous, but they never knew the reasons that was source of this nervous situation.

I am telling you now: Sunfish has polarization sensitive vision not found in humans and other vertebrates mostly which helps in enhancement of visibility of target objects about scattering media, or using a method that called polarization difference imaging. This sunfish is considered as an invasive species in the sunshine place of Florida so habitants in that place never can use for exhibition or research or Aquarium without permission of some agencies as they are naming over there for while those habitants will take punishment for illegal action of possess Sunfish, so it is so important for science this Sunfish, and we are proud for our Sunfish too.

Not only you but and all of us said very nervous this time Blue fish! All fishes turned their head in its direction while they moved left and right their tale while the water started to go up by this suddenly strong moving. Yes I am very right because is and our fish this Sun fish because it is with some blue color in its body, but this stranger fish is not telling its appearance. At that time another fish spoke loudly: this Sun fish is and our fish and we are proud for that fish, because in tis body it has and some yellow color in tis back and sides also has some yellow scales too. When Juvenile Pike Esox Lucius saw that was started one unaccepted conflict said with loud voice: Please leave me to continue my presentation because everything is coming in line, so it is art of my speech but I do not know why you do not have patience this is Science of Convention and we need some specific details to explain so give me time to speak.

At that time all fishes got calm and the Yellow and Blue fish they saw each other and went back to their place. While Juvenile Pike Esox Lucius continued its passionate speech. As I make remarkable that our Green

Sunfish, I am saying that because this fish is classified by Science fishes's board and by Science's habitants of Earth in our family but I am explaining to all of you that this Sunfish has blue color too in its body and yellow color too, it is with big mouth and typical length is going 3- 7 inches and weighs less than a pound.

The maximum length of the Sunfish is going to 30 cm or 7 inches. With maximum weigh of 960 gr or 2, 2 Lbs. Identification of the Sunfish by other species is so difficult to do because those species hybridize. Our green fishes are all around this planet like in those places that named North America, Rocky Mountains, Hudson bay to Canada, Gulf Coast, Northern Mexioc, Mississipi River, while Sunfish is transplanted to Europe water, Africa and Asia too. Our Green fishes and Sunfish too, like to hidden it selves around the rock and other objects that are providing cover and protection, also like to live and to vegetate in lake, sluggish backwater to pond s with gravel, sand or bed rock bottoms. Green fishes are able to tolerate poor water too. Our green fishes are feeding themselves with larvae, insects that fall into the water, small fishes that is not good thing, also and invertebrates too.

At that time one fish from audience asked loudly: So you want to say that Green fishes are aggressive fishes.

No answered strict "Juve – Lucius", I am saying that some times in difficult condition they are using small fishes. So "Juve - Lucius" wanted to make smooth and to correct its speech and those facts. Our male green fishes are using force to protect it selves by other males- fishes. About reproducing, our green female fishes are dancing and swimming together with male green fishes until they (female descend) are deposit the eggs on the nest of male green fish that will lay from 2000 to 26,000 eggs and leave them to male to guard. Green male fish keeps watch those eggs until they hatch in three to five days, while are protecting and fanning them with its fins. Green male fish is keeping those eggs clean and is providing them with oxygenated water. After the time that eggs get hatched, the male will often seek to attract another female to lay her eggs in his nest. What is more important that I need to explain to this Science's Convention about our Green fishes' reproducing, is this fact that is so valuable for the science.

Our females Green fishes and specific Green Sunfish tend to nest those eggs not only to Sunfish male but and to others different species, this phenomenon will lead to the next generation to face some amount of hybrids. These Green Sunfish hybrids will often look like one combination of their parents that will make so much difficult to distinguish one species from another. This is problem that this Sunfish has so many different colors in its body like blue, yellow and some dark spot, but dominate the green color, this is one big point with high value for Genetic science, this is the future because this Sunfish is so beautiful too, with this combination but it is in our green fishes' family this is the true. So I am sure that I make clear for those two fishes blue and yellow that opposed me before some minutes ago.

The Sunfish science name Lepomis (Scale Cover) is coming from one place name Greek on Earth while Epithet Cyanellus is coming from same place Greek that means blue.

All we Green Fishes are peaceful fishes, because of our some green hybrid fishes we have so good relationship with some others fishes family like Yellow, Blue and some others too, so we want to swim and live in peace with all others fishes of water and for our good looking and our good feelings for others fishes - species we are giving one big thank you to our honor "Water - God" Poseidon.

At that time all other fishes gave one strong applause while the surgeon Fish came at front of stage and kissed Green beautiful "Juve – Lucius" while its tears were going down and to be members of big blue ocean's water while one big enthusiasm was creating around by its very good speech so all forget the interruption of the jealous blue and yellow fishes too. Two blue fishes said to each other, now Is coming our blue fishes's family for speech, strangely everything is going so beautiful like magical in this Science's Convention, we hope our family to shine too to be bright with our representative. At same time all others fishes started to swim and to speak loudly enthusiast with full happiness for this big event while were doing so many different comments. Looked that Water's God Poseidon has used all its talent to throw out on this blue water different beautiful colors about painting those different form of those beautiful fishes.

Chapter 11

Blue Fish's speech!

Surgeon fish started to explain to others fishes, that another blue fish wants to speak on behalf of Blue fishes community really are so many requests from this community of Blue fishes, like Blue Parrot fish, Blue Dragon Fish. Blue Marlin Fish, Blue Bright Fish, Blue Transparent Fish, Blue Art Fish and Red Blue Fish. I am thinking that this community has so many energize fishes and those wants to speak up with you, so I am giving them opportunity to speak but with limit time. All others fishes gave one strong uauaua, while some of them were speaking, that they never knew that will have opportunity to speak and others in this convention, only one representative, so was created one noise.

At that time Surgeon fish said to fishes in audience:

I want to explain to you that is not my mistake but is problem of information that your leaders gave to you, or the way how they interpreted to you, while about Blue fishes community they have sent their requests for some fishes that wanted to speak one week before Convention, they sent their requests with emergency by one ray gas that was moving and diffusing by electrons all molecules of water in form of spirals and their message came with speed to me,.. So they have done their job about request and I am doing my job about approving their speech but with one condition their speech will be only 3 (three) minutes for everyone. This condition is to give them opportunity that all those to speak also to see their speed of their eloquence of speech and concentration of their topic. After its explanation Surgeon fish called name of Blue Parrot.

Blue Parrot fish came fast and started its speech. Our Blue fishes community is coming from family of Pomatimodae with science name Pomatomas Saltratrix, as you know we are all around this planet while we are knowing to Australia like Tailor, to East coast of Africa like Shad and Elf on the West coast. Other common name of us are Blue, Chopper, and Anchoa. Our blue color is dominating through our body. We have length seven inch or 18 cm with weighing that is going to 40 pounds (18 kg, but generally average our blue fishes weighing is 20 pounds or 9 kg. Our blue fishes are emigrants in different seasons of the year. We are everywhere in tropical and subtropical water. In winter time our blue fishes are in Florida sunshine place, through April they are swimming in North in June blue fishes are going to one important historic and beautiful place that is named Massachusetts and Nova Scotia, by October they leave the waters around New York and are heading to Gulf of Mexico.

Our first cousins that are leaving in Europe are emigrating during this period time after October to Black sea in Europe place, also to one very beautiful; and planet historic' s place as are named habitants of Earth that place Istanbul specific in Bosphorus, Sea of Marmara, Dardanelle, Aegean Sea and all around Turkey coast. Also they are swimming with happiness to one very warm place named South Africa. I want to make bold my phrase to all of you now that we beautiful Blue fishes are everywhere in Europe, America, South Asia Africa Australia, New Zeland. I am very proud for my blue fish family and for our peaceful spirit. I want to give time to my friends Blue fishes so I gave in general way about my community while I am saying to you thank you for your listening. Applause came strong through water by all fishes.

Blue Betta Fish was swimming quiet with one big smile started its speech: As you have heard about some well-known name about exact science of this planet about Alfa and Betta that those term are using everywhere, really I am Blue Beat fish that really I am in group with Science's group of fishes. We are doing different experiments about using different nutrition for our fishes also we have connection with all others objects of this planet that have in their name this common middle name Betta, for example we are using some concentrate of Sugar Betta or Sugar

Beet that have different name, their concentrate with sugar we are using in some water when salt is in high level so we are doing neutralize of the water for our fishes. I came in this convention to tell you about this specific job of neutralize water that is science also to tell you that I am proud that I am from Betta family of Blue fishes. Thank you very much for your attention. While one fish from audience said good job this experiment to make neutral salt water. Thank you Betta fish for this experiment.

After Betta fish came to stage Blue Dragon fish that its name was well know all around the blue water. All fishes started to move to see better this fish. Blue Dragon Fish came on stage with its very beautiful and very characteristic form that gave wonder to all other fishes. This fish said with speed its phrase: I am Blue Dragon fish well known for my beauty and for my dragon action that is my exception point. Our family's fishes has big eyes and big mouth too, they have really fear reputation but is not like that. Our fishes are eating in darkness their prey, they are laying 30 to 90 eggs and are getting maturity after 3 to 4 years our length is to 20 inches. Longer length of our fishes is 90 cm or 35 inch, and are staying in deep water until to 1500 meters.

One very important point of our Family Dragon fishes is the Photophore that we have behind of our eye that is acting like headlight during time of hunting of pray during darkness. This is in need to be used by science, so we have and something good very specific for future science is not only fear reputation that are spreading for us all around. Our family has name Stomidae, and we are living more in Pacific Ocean. Thank you very much for your attention. At that time all fishes started to speak with each other for this curiosity that they never knew before. At that time one very elongated fish with its longer forward dorsal fin that is created crest, came to stage and said I am Marling fish form Istiphoridae's family. We are sport fish and we are swimming so fast about 80 km /hour or 50 mph. We have our members in Atlantic are with length 5 m or 16, 4 ft, and 818kg or 1803lbs that has name Makaira Nigricans also Istiompax Indica which can reach length of 5 m or 16, 4 ft and 670 kg or 1, 480 lbs in weight. Really those two fishes are sport fishes, I came here to tell you that our body need to attract Science's fishes to study our build body and to create others sport fishes to swim fast and to save it selves by predators fishes. Some fishes said this is very great point for future science and for all of us that we are facing danger all the time by wilder. At last Marlin fish said thank you very much that you listened me.

Blue Bright fish and Blue Transparent fish came together at stage while with one big smile they said in same time: Hello everyone by both of us Blue bright and Transparent fishes. All gave one strong applause. Blue bright said I am here with my lover blue transparent fish to tell you that we are for Peace and for bright life in our community and other species' communities. We both are studying one experiment about Fish brain that so many habitant in this planet are skeptical and sometime sorry that I need to make bold this phrase but they are humiliating us when they want to put down each other are saying "Fish Brain" really they do not know how many value points all of us different fishes have for science. So we are studying the brain of fish how to make to absorb more bright and transparent every phenomenon in the water and behavior by others. At that time Blue Transparent fish said we are working and studying hard but we have so many obstacles by Blue Devil fish that it is in this convention here, so we need by this convention to take action to this Blue Devil fish and to create us to continue in calm situation our studying and experiment that will be so good for all of us and for future.

One very beautiful purple fish asked with high voice: please tell us what is doing to both of you this stranger Blue Devil fish?

This Blue devil fish, is undermining our job, is erasing our writing with ink in water. When we saw this phenomenon we did not know who was doing when we discovered we decided the result of our experiment to write to ice in very low temperature so no one will destroy our writing but after some days when we went over there to write other results we saw all our writing was erased, because the thick ice was broke with some strong objects as we understood. So we decided to interrupt our experiment and our writing until this event to happen so we need your helping for that reason we came here.

Some fishes spoke loudly: Blue Devil Fish not to stay anymore in this Science's Convention. All others fishes approved with one voice: Blue Devil fish not to stay in this Science's Convention. At that time Blue Devil Fish that was swimming in hidden way whispered with itself: I do not like this couple what they are thinking that are!, but it left this Science's Convention because it saw all were blaming it. After that situation that fishes finished their yelling and screaming this couple Blue Bright and Transparent fishes said: Thank you very much for your attention to our speech.

Another Couple was swimming to stage with full happiness while they started their introduction: I am Blue Art Fish, while this is my very lovely friend red Blue Neon Fish. I personally came here to discuss for all of us for future. As you know now I am Blue Art fish, I love art, I love peace but I see so many aggressive action in blue water by different wilder. As you know all Blue fishes are preyed by larger predators at all stages of their life. Some of those predators are larger Blue fish, Weak fish, Tuna, Sharks, Rays and Dolphin that are danger for Juvenile blue fish while for adult blue fishes are, Billfish, seals, Sea Lions, Dolphins Tuna, Sharks too and some others species. We need to be together and to build one new protection's system for all of us and to defeat those wilders. I want all youngest fishes to love and to participate in art. Thank you very much for your listening of my speech.

While Red Blue Neon Fish was swimming more close to Blue Art fish and said with smile: I came here to give some love to all of you with my beautiful color, that is dominating red but I have something blue and neon color. My red color is lovely color as said some habitants in one small Mediterranean place that named Arberia or Albania that red color is color of love, while neon color is helping me to shine the way that I am swimming, so I want from this Convention and from all of those beautiful lovely fishes one strong cheer for Love. At that time looked that one strong force moved the water that its waves were going to meet the sky from moving of Fishes with happiness.

All they screamed with one loudly voice: For Love for love. While the blue water that got form like water fall got multicolor dimensions by reflection by the Sun's rays of all colorful fishes. It was created one magnificence view inside this clear blue Ocean's water.

Through this happiness this couple of fishes said: Thank you for your attention and for your support of us. Water - God bless this Science of Convention.

Chapter 12

Whales

Whale's speech

Surgeon fish came to stage while started to speak and to give break time for that day convention one terrible noise came all around and all beautiful colorful fishes, got wondering and scare. No one knew what this unaccepted noise was. At that time they saw that one big whale started to surface above water while started with big noise and loudly to speak.

You never can do this Science's Convention without our Whales's family presence. We are big Whales with big biography and history for his planet and history of water. Our existence has time since 54 million years, plus we are coming from Ceaceans family so I need to speak in this Science's Convention, I need time for my speech. Our origin name is coming by old English Hwael also from Proto Germanic Hwalaz that is common of various marine mammals of Ceacea family.

At that time all fishes were confuse and they start to speak with each other what to do now. Most of them started to show nervous and scare feelings. At that time Surgeon fish that was very smart turned its head to all audience's fishes and said I am allowing this Whale to speak over there in that place.. but Whale said: I want to speak in stage in front of all. Surgean fish replied immediately: Our stage is done with ice, that we worked to make this form frozen until – 50 grade under below zero, that in that way this stage to afford warm weather for some days, but this stage can't afford your big weight so can get broke we do not need that, but we are doing that thing all fishes to stand straight in your direction and you will be in front of them. All fishes approved and Whale looked that accepted that version by very smart Surgeon fish that did not want to give so much honor to this scare Whale for all fishes. So its diplomacy resolved this difficult situation for both sides, whale and small very beautiful colorful fishes too.

At that time one yellow Chichlid fish said to another Blue fish: do you listen our very smart Professor Doctor Surgeon fish, but it is very smart because it has PHD degree. While Blue fish said: I have Master degree but never I will find so fast at that moment this good solution in that difficult situation. Ohhh you have achiever so many goals in your life but I know some others fishes with Master degree that they never achieved anything in their life and why they have high degree, but really this PHD degree is so good for our Free Fishes.

After that discussion all fishes got position in front of big whales, while that whale got pomposity position and show so much its big body in front of them. At that time some small youngest Whales came and stood behind big Whale with big smile. At that time Whale started its speech. Hello every one of you all beautiful fishes. I heard about this Science's Convention but really I did not know the time and day, so I was controlling all the time around until I found this place with all of you. Really I am with broken heart that no one gave news to me but now I am happy to attend this Science's Convention.

I have some problems to discuss in this Convention: As you know we have big body that is going to 30m (98 ft) with weight 30 to 160 Tons or (160.000 kg) for blue whales while the small whales to Pygmy species whales to 3, 5 m (11ft). Our longer life is going until to 77 years, bowhead Whales are living until to 100 years one century, but bowhead of Alaska one place on this planet is living to 115 until to 130 years,. One curiosity that you need to know is that "Aspartic Acid Racemization in whale eye combined with harpoon fragment indicated at age 211 years that is doing this specie bowhead whale the lonest live mammal, while we are 407 diversity species all around planet's water you know we have close relative Hippopotami. Family Ceraceans and artiodactyls classified under the super - order Cetartiodactyla which includes both whales and Hipopotami.

All of us whales, Dolphins and other branches of our family during time we have got so many transformation depended by environment. We are feeding our youngest with milk from mammary glands, we have warm blood, we breath air like all other mammals and we have body hair. Beneath our skin lies one layer of fat that is called blubber. This blubber is giving us energy and insulated our body. The neck vertebrae is helping us to be flexible and to get adaption during our swimming, so our Neck vertebrae is one important point for This Science's Convention.

Whales breathe by blowholes, Baleen whales have two and Toothed whales have one, these are located on the top of the head, that are performing completely breathing. The earhave specific adaption to whale while those are taking sound by throat.

Our males are called bull and female are called cow while all new born are called Calves. Whales do not have fixed reproductive partnership. Female are feeding with fat milk new born into their mouth. This feeding and taking care for Calve by cow mother for new born baby is continuing for one year. Re productivity is occurring seven to ten years.

One very important value of whales is that those are able to teach to learn to cooperate, to scheme and grieve because the whales have spindle neurons in area of brain that are homologous to humans so suggestion is that those neurons perform a similar function with humans that are habitants of this earth planet. So our specific spindle neurons homologue with habitant is one big point of science research I think, remarked blue whale while was seeing all round with full pride, and was absorbing the reaction of audience's fishes.

At that time Blue Whale heard one strong reaction and discussion of fishes with each other with enthusiasm for this big news. Whale at that time said: Wait I have another big news for this Science's Convention:

Unlike all other animals Whales are conscious breathers. All mammals sleep but Whales can't afford to become unconscious for a long time. So I am telling you right now that Whales are sleeping with one side of their brain at time they may swim, breath consciously, also a void both predators and social contact in water during this period of rest or sleeping as we naming it. So this is another big news to study our brain about this two functional or so many functional during rest time for Science's Convention.

At that time the Whale heard one strong reaction by all like uuuaaaaauuuuu. We never knew about that.

Whale continued its speech. Most of the time Whales are exposing their body on surface of water also for their sounding Whales are staying close to surface of water.

One specific Whale specie that is humpback whale are communicating with each other by sounds that other are naming those Whales' sounds, so this is another news for this Science's Convention. Without vocalization by sound we are doing some functions like echolocation, matting and identification this is very important for science in this current time for all fishes.

We are known like mimic human speech that is telling to others fishes our strong desire to communicate with humans or habitants of this planet. Really we have different vocal mechanism so to speak like humans or habitants of this planet are taking one strong considerable effort.

We are so important because others and scientist habitants of this planet called us all Whales like "Marine Ecosystem Engineers". At that time one green fish spoke loudly: As you know you are called predators by so many because you are eating small fishes until to big,.. at that time Whale that was prepared for that situation said: is relative how you understand our feeding but this is the law of survive, we need food to keep our big body but I am telling you how many positive points we have for all fishes in this water.. you know predators are not called only Whales or Baleen but so many others small fishes too that looks so beautiful.

We are like reservoirs of nutrients, such iron and nitrogen while we recycling them horizontally and vertically in water column.

We are providing energy and habitat for deep sea organisms.

We are so positive for influencing of productivity of ocean fisheries, while other are termed us like" Whales Pump

I have one big problem now for this Convention that is very immediate for our existence: Habitants of this planet are hunting Whales for their oil, so our numbers is reducing in maximum, this phenomenon started since the seventeenth century and is continuing until now.

We have 10 main different species of Baleen Whales that are:

1- Blue Whales or Balanoptera musculus.

The Blue whales is the largest animal that ever lived in planet of earth there are much larger than the largest dinosaur. So if habitants are using different specific places to honor dinosaurs in their museums why not for us, also if they are doing so many movies for dinosaurs why not for us, but this is problem of them not of this science convention but I want to make remarkable about our magnificience. Only for curiosity I am telling to you one fact that I have studied statistic in our Fisher's University that was written in one Blue book that between 1930 and 1971 about 280.000 Blue Whales have been killed. This is big tragedy for our family. Since 1966 not many Blue Whales were left. The Baleen that is in this family is feeding in summer in north polar waters while in winter in subtropical or tropical water. They are swimming in group 2 or three animals. Really all Blue Whales are 11. 300 worldwide. They are 24 – 27 m with weight 10 -120 tons

2- Bowhead Whales. Balena Mysticetus.

This Whale is going with length 14 -18 m female are more larger than males while average weight is 60 – 90.000 kg. Bowhead is blue black color with big head with smooth skin and free of parasites. Located only to four area that are Spitzbergen, Hudson bay/Davis strait, Okhost Sea and Bearing Chuckchi/ Beautfort sea. Bowhead is seen alone or in group with three animals. It feeds generally with small crustaceans.

3- Bryde whale. Balaenoptera Edeni.

Bryde's whale length is 11.5 – 14.5 m with weight is 10 – 20.000 kg. Bryde's whale is 20.000 – 30.000 animals North Pacific. Bryde's Whale is dark grey on the back and lighter on the belly.

4- Fin Whales. Balaenoptera Physalus.

The fin Whales have length 18 – 22 m sometimes is going to 27 m with weight 30 – 75.000kg. The females are more larger than males. It swims in group with 10 animals. There are 100.000 fin whales in Southern hemisphere but only 30.000 in the northern hemisphere. The fin Whale is a fast swimmer with speed 30 km/hr.

Its color is dark grey to brown, but more dark is going down left side.

5- **Gray Whales Eschritius Robustus**

Gray Whales are with 13,5 to15m in length and weighs up to 27.000 kg. During migration females with calves swim with 6 adult. The Eastern population of gray Whales is 21.000 animals while in Pacific Ocean is small probably 100 – 200. They color is gray all over. The gray Whales are spending their winter time in San Ignacio Iagoon Baja California and Mexico places of this planet. They are organized and sometimes they are going very close to the habitants' boats to give them pleasure so they can touch their smooth skin. In summer they are traveling to another place in North of this planet or West - North that named Alaska because over there is more cold and more fresh air

6- **Humpback Whales Megaptera Novaeangliae**

Our Humpback Whales or Baleen as are calling habitants of this Earth planet is well known like Whale of songs, the function of songs no one can understand is big mystery of our Whales is secret in our family. These songs can be heard in long distance. The winter the Humpback Whales are spending in Hawaii and Baja California places while in summer are going to Alaska place. There are 19 m with weigh 48.000 kg. They are black sometimes is going black to white color. They are swimming in group with 3- 4 animals, while are located in three specific area in North Pacific, North Atlantic, and Southern Hemisphere. There are 5,500 Humpback in the North Atlantic, 2500 in the North Pacific and about 12.000 in the Southern Hemisphere. There are about 500 these kind of Whales in Indian Ocean.

7- **Minke Whales Balaenoptera acutorostrata**

The Minke Whale is 7 to 10 m with weigh 4,500 to 9.000 kg. The back of mink Whale is black and belly region is white. There are

18.000 – 27.000in the North Pacific place, in Northeastern Atlantic: 90 – 135.000 in the Central Atlantic about 60.000 and in Southern Hemisphere 200 -400.000. Compare with others Whale the Minks Whale are multiply while others are reduced.

8- Pygmy Right Whale. Caperea Marginata.

The Pygmy Right Whale is the smallest Baleen Whale species. It is not closely related to the Right Whales. The name comes from its appearance, mainly the shape of the mouth. Very little is known about this species. Since it is small and usually swims alone it is very difficult to find in the open sea. Most data is derived from dead stranded animals.[48] Its length is 5.5-6.5 m. It weighs 3,000-4,000 kg. The lower jaw is bowed and protrudes slightly. The body is stocky. They have a dark-colored back, which becomes darker with age, and a pale belly. This whale has a prominent dorsal fin. The flippers are small and rounded and located under the body. The tail flukes are broad. The Baleen plates are yellowish and are up to 70 cm in length. It is only known from the Southern hemisphere and usually is solitary. Diet is not well known, but probably plankton.[42]

9- Right Whales Eubalaena Glacias (Northern) Eubalaena Australis (Southern.)

This name Opur baleen whales has enjoyed by Earth Habitant because in their language they are saying they right to catch. Right Whales are slow and have a lot fat also have length 11 -18 m long with weighs 30 – 75.000 kg. they are black and really very, very fat. The Right Whales are swimming comfortable to Antarctic to Australia, South America,, into Indian Ocean, north Pacific from one place that name Japan and Baja California. There are about 1.000 right whales in Northwest Atlantic and 1.500 in in southern.

10- Sei Whales Balaenoptera Borealis

The Sei - Whales are the fastest swimmer among the Baleen Whales. It can swim to 38 /hr. Males of Sei Whales are 12 -18 m, 22, 000 kg, while female are 20 m long with weigh to 22.000 kg. its color is black while the chin, throat and belly are white. There are about 13.000 in North Pacific, in Antarctic about 40.000. One fact by Fishes Statistic's University is telling us that the annual catch of Sei Whales in 1960 was 10-15.000 per year. Hunting of Sei Whales was stopped at 1970. Thank to our Water God Poseidon, for that reducing of this hunting.

I am speaking so short for our Baleen Whales that are well known but I do not want to make tired you about 26 species that we have with others Whales like toothed whales, Dolphin and Porpoises. I decided with courage to stay and to insist to speak to all of you about our big value that we have in this Planet and more important about some very important point that we have about new research of this Science's Convention plus we have and so many value's point for art and movie that are using for pleasure or entertainment the Earth's habitants.

At that time all beautiful colorful fishes forget their scare but made one ooooo we never knew about all those characteristic of Whales. They were shocked by this information. One Red fish spoke to one Blue Star fish and Red star fish: Okay I heard their big value but this does not mean that we do not need to be warning for them anyway they are so big and they are not familiar with us plus they are in some aspects so danger for us. The stars did not speak but they were thinking how to turn in value process the positive points of Whales and why the small Red fish was insisting with its speech to convince their mind. At that time Surgeon fish spoke without expression in its face but was moving its tale: Okay we got so many good news today and we need to make research about them was good information for us and helpful.

I am saying to Whale thank you very much for your speech, now we need all of us to take break time and we will start again after some times when we will feel that the water will come more cold at after noon, so we will be more fresh in our thoughts and discussion, at that time all colorful fishes gave one strong applause for Whale's speech. After that all fishes were enjoying one suddenly show that four youngest Whales were doing around the big whales only to give greetings and one big thank you to this audience of fishes. The youngest Whales were so happy and started to jump on surface of water while all fishes were happy for that show and were laughing too, with their jumping up and down in blue water. Surgeon fish whispered to itself: thank Water God, thanks Poseidon, for this good end of Whale's speech and Free small colorful fishes too. Is good that scare left it place to this enthusiasm. Thanks my water God Poseidon.

Chapter 13

Hawksbills Turtle | Sea Turtles's speech

After short break all fishes were swimming back to convention's place. At that time one huge crew of yellow Chichlid fishes came fast in front of stage while behind them some others yellow fishes that were swimming around the one Sea Turtle, were not allowing others fishes to become close with this turtle. The Sea turtle was swimming so slowly while looked that did not care about others fishes that started to whisper with each other. One big yellow fish was swimming straight to Surgeon Fish's place and started to move its arms on water while was speaking with Surgeon Fish. The others fishes saw one gesture of Surgeon Fish that made them to understand about one approve action.

The yellow fish was swimming back to Sea Turtle's place and was explained to it about conversation with Surgeon Fish.

All others fishes were swimming in line and took their place to attend this convention. All around the sharks started to swim nervous and were making so much noise but no one did not understand why was this big noise. When all fishes stand to their place the Surgeon Fish got to stage and invited the Sea turtle to come to stage. The Sea Turtle was swimming straight to stage at that time one water fall was created by nervous sharks that scared all fishes around. Sharks'noise was going out of control. Whale from distance shouted water up and started to jump in water while said:

Do not care for them, they are jealous for me and for Sea Turtle's speech now.

One shark was swimming all around while was moving some water with speed and said of course we are claiming for our speech in this Convention because we are water's habitants too, we are in water world, and we have our value for water's world, but now we are not jealous but we are wild nervous, for all of yours'skepticism about us.

You are wild and danger for all of us spoke and yelled one blue fish.

Shark nervous replied stop your fsfsfsfs, you are nothing compare with us, but you know one thing only to make mixer others, stop your fsfsfsfs Blue fish because as is your color, is your heart's color too, stop your fsfsfsfs.

Surgeon Fish saw this scenario and interrupted Sharks and Blue fish with its speech. I am telling to all of you this is Science's Convention, of course we need to know about each other and to give information to all others fishes, but our science is for life so some water's habitants or animals have limitation about this Convention for their wild nature. We are doing this Convention to make more longer life for our free fishes, to make more improvement for that life and to create security for so many small fishes by wilder, so with my respect for Poseidon God that created you shark and made possible your existence in our water world, I am telling to you I am respecting our honor Poseidon God about you but I am not affording your very wild nature, that you are so greedy all of you sharks, younger and adult to make danger life until to mortality about free fishes of water's world and planet's habitants or human as they are naming themselves, so I can't allow you to get honor in this Science's Convention of life, this is for life. You must to say to me one big thank you that I allowed all of you to be here close with us. I am clear with you now.

No replied Shark you are not clear because our history is older than your existence our Fosiles are giving us pride for our longer existence.

At that time Surgeon Fish said to Shark:

Go to find and to study your Fosiles and write your history or do your Convention of Science with your wilder sharks and all others wilders of this water world but not here in our peace Science's Convention.

I never will allow myself to give you this very

"Luxury Opportunity" to speak in our Convention. You can do what you want if you are lifting all this water like one big water fall I am not allowing you to speak in this Convention, because this Convention I created not you.

All other sharks gave one big noise with one shouting voice with one strong Oooouuuu.

Ouuuu and uuuuuu never I will allow you and never withyou closed its speech Surgeon Fish.

All colorful fishes got freely breath and gave one strong applause.

The whale started to make some beautiful jumping in water with full happiness and all fishes continued applauses.

One big shark was swimming close to Surgeon Fish and started to speak very nervous:

So you are sending us out of this Science's Convention?

I am not allowing you to speak in this Convention, but if you want to listen you can stay over there behind small colorful fishes, if you do not want is your choice what to do to stay and to listen from peace fishes good news or information or you can leave this place of Science's Convention and to go to your place between wilders, it is your choice and yours others sharks choice too.

Sharks did not speak anymore but made one strong tuuuurffff, tuuurtfff, in water while with its tale started to move left and right water. All were seeing this shark and all others youngest sharks that were swimming behind it, but when the big shark stood behind big crew of the small beautiful colorful fishes and made with sign to its youngest sharks to be quiet, all got calm. At that time Surgeon Fish was swimming back to the stage and invited the Sea Turtle to stage.

All small fishes swimmed fast to stage to open way for Sea Turtle while all others were swimming aorund the Sea Turtle until the Turtle arrived slowly to stage. SeaTurtle created one good impression with its beautiful cover bold brown color and some yellow stripes that were creating some square figures on its cover while its down of body was simple yellow. Sea Turtle was proud for its beauty that Water God Poseidon has given present to it.

Sea Turtle.

Turtle started its speech: My name is Hawksbills Turtle. We have value about our contsruction of body and for our very interesing color that we have in our body too, as you know our honor Water God "Poseidon" gave this present to us and made us valuable. Our name is because of our narrow, pointed beak, we also have a distinctive pattern of overlapping scales on our shells that are making us valuable on market like" Torteiseshell"

Our Family's Turtles are swimming everywhere in tropical Oceans, while those turtles are predominantly in coral reef. Our Turtles are the living representative of reptiles that has existed on Earth and travelled our seas for last 100 million years. This is one information for that Science's Convention. Our turtles are link in marime ecosystem and are helping the health of coral reefs and sea grass beds. Turtles are eating jellyfish, and sea anemones, is another way of their feed that turtles feed mainly on sponges while are using their narrow pointed beak to extract them crevices on the reef. This si really for science study.

All around the Earth and this planet are working habitants with fisheries to save us because by so many reasons our numbers is reduced so much that is one big tragedy for our specie. Habitants of Earth planet have created the most sofisticated equipments that they named Satellite for following us and saving us by illegals habitnats that are using us for their greedy feeling for big profit also those Satellite are controlling our migration to save our Turtle specie also are folloring our moving place to place for that fact we are greatful.

So we are proud for our value and we are giving one big thank you to those Earth's Habitants that are protecting our specie, all our beautiful Turtles. You will say that we are eating jelly fish but sometimes we are doing that in very difficult time for survive as every animal but really we are not proud for that. I came here in this Convention for that important reason that this Science of Convention to create one new food for us to stabilize our food and feeding so we do not want to eat jelly fish anymore, because we do not want to be aggressive really si not our nature.

My Turtles'family has another big problem because we need to reach surface to breath, that this situation is bringing for us so many dangers by others animals. I am doing one request to this Science's Convention to create one supply for us that to make more easy our breath without reaching surface, this is important for us to survive and to make quiet life, to avoid different dangers. I am giving one idea one novation for example if this Convention with high level of Science's Fishes to create for us some longer tubes to be connected with our body and to go up above water, this will help us for our freely breathing.

One red fish asked: What is your suggestion about what material to prepare this tube?

You have very interesting question you are right Red fish but I have one idea to help Scientist's fishes. This tube maybe must to prepare by flora or grass under the Sea or Ocean to make mixer with some materials of dead Octopods that are transparent and to melt both of them while to put that material to one dead body of longer fish and to put on ice after to send out of this body and to connect with our body this is one novation but Scientists' fishes can create something more sophisticated and soft for our body.

All fishes gave strong applause with full happiness. With our swimming through the Tropical Ocean's water with our beautiful brown and yellow carapace plates makes our tortoiseshell like ornament items in different stores of tropical places or for jewelry, for Earth's habitants that are coming for vacation and fun on those tropical places. On Earth Planet in one place that named Asia we have domination about those ornaments or jewelry with our Tortoiseshells.

God Poseidon blesses all of you Turtles started to say with high voice so many colorful small fishes.

While Surgeon Fish with one strong huge to Turtle said: Thank you very much Dear Sea Turtle from Hawksbills family, thank you very much for your information and for your great request. I am proud for you that want to save your community and your specie that is very important in marine ecosystem.

All others fishes moved around and started their strong applause while said very good, very good, beautiful Sea Turtle, Hawksbills Turtle, we wish that:

Water - God Poseidon bless you.

Chapter 14

Star Fish's speech!

The Surgeon Fish came at the stage while gave one thank you to Turtle started to make some comments about those very beautiful high level speech of different colorful fishes and those great information of this Science's Convention but at that time one very beautiful Red Star fish started to say I want to come to speak in stage because you never know information about our great beautiful family and how important are specific us for this Science's Convention.

Surgeon fish saw with full attention and strangely and suddenly asked with high voice all colorful fishes in audience – water? What do you think all of you very beautiful colorful fishes that are so lovely to me, to allow this Red Star to speak in this Science's Convention or not?

But before to get answer by all others Audience's fishes Surgeon Fish heard again Red Star that said fast. I need to come with some of my stars fish in that stage to show so many things our beauty our diversity and our value for science in this "Bios – Water"!

At that time all fishes said with one voice allow Red Star fish to come with others stars fish and to speak on stage. Allow it. Allow to speak! Surgeon fish laughed after it invited Red Star fish to come with its favor crew on stage.

Suddenly one melodious voice came out of mouth of Red Star Fish that gave big surprise and wonder to all colorful fishes.

So the Red Star Fish was swimming with its favor song to stage:

I am very beautiful Red Star,
that I am swimming through,
I am the most happy Star fish in this water blue.
I am enjoying all this Water - Kingdom
with my lovely colorful stars too,
That our honor king Poseidon
gave like value present this water blue.
Uuuu huhuuu uuuu huuhuuu (was making only melodies.)
I am happy for my beauty,
while I said to God Poseidon thank you,
I love all my beautiful stars friends
while I am proud for their beauty too,
I am saying always to god Poseidon Thank You.
Uuuu hhuuu uuuu huuhuuu
I am happy Red Star
I wish that leader to be
for all my colorful stars fishes
in this water blue,
I want with my five red arms
to spread all around this "Bios - Water" World through,
Love, Peace, and Beauty in this water blue,
Uuuu huhuuu uuuu huuhuuu
I wish behind my five red lovely arms
to come all others fishes colorful
with their diversity
to make beauty this "Bios - Water - Blue"
and to give to our honor God Poseidon
one big Thank You.
I am happy beauty Red Star,
I am dedicate for Love Peace and Beauty of
Bios -Water - Blue,
I want to be leader of you,

Uuuu huhuuu uuuu huuhuuu
My five beautiful red arms
are segments – connection
of Universe's rays full radioactivity
to give full happiness and prosperity
to all Continents of this planets and
their all "Bios – Water - Blue" in infinity.
Hiii hihiiii iiii hiihiii
I am Red Star I want Prosperity, Peace and Love,
I want to be leader of all of you while
I am saying to Honor god Poseidon
One Big Thank You.
Uuuu huhuuu uuuu huuhuuu

When the Red Star Fish finished its song and arrived to the stage was created one very strange situation by all colorful fishes that gave one strong cheer, applause, screaming and whistles in different voice and melodies, so the red star did not start its speech until to finish its loudly noise all free colorful fishes, until this big enthusiasm to come more down. As was seeing by all fishes enthusiasm will continue for so long but Surgeon Fish called from stage all of them to calm down and to hear information and speech of Red Star Fish. So all this beautiful crowd that had so much respect for that Science's Convention's leader came calm.

Surgeon Fish said to crowd: Now in front of you is the beautiful Red Star Fish that gave one big pleasure with its song.

After Surgeon Fish asked Red Star:

Who created this beautiful song dear Red Star Fish?

I have created?

Who gave to you inspiration for that song dear Red Star? …asked Again Surgeon Fish.

My inspiration was all this present by our honor God Poseidon, gave to us, this "Bios – Water- Blue" look around all those beautiful colorful free fishes, also during my swimming I saw one full diversity of Water – World, with fauna and flora so I got inspiration for all this resource in our water world for good life.

So what you hope or you wish?

I hope, I wish and I want to achieve Prosperity, Peace and Love for all. All gave one strong applause.

Surgeon Fish said I want to follow and to be part of your inspiration.

At that time Red Star fish called some its Star fish, like Blue Star, Orange Star, Green Star, Purple Star

Pink Star and Royal Star Fish. All were swimming in line straight to stage.

When all were stars fishes were on staggered Star started its speech: I am Red Star fish from Asterinidae family. Our science name is Callopatiria Granifera. We live all around but we are dominating to Namibia place and South Africa. Our food are detritius. We have 1500 species that are swimming from tropical water of Ocean to Frigid polar waters of Ocean. Our star fishes are living below water's surface to intertidal zone down to abyssal depth to 6000 m (20,000ft).

We have some characteristic like our mouth is in center in lower surface, also we have one "tube feet" operated by Hydraulic system that is value for science. Some fishes are naming us predators on benthic invertebrates. We can reproduce both sexually and asexually this fact I think is so much value for science too. Another value point of our family is because Star fish can regenerate damaged parts of lost arms and we can shed arms like defense. So many stars fishes are brightly colored in various shades of red or orange, while others are blue, grey or brown.

All starfishes are "Marine invertebrates". Supposed our existence based on fossil that are ancient that dating back to the Ordovician, found by science habitant of earth and our science fishes that the history of existence is about 450 million years ago. We stars fishes have five arms but some have seven arms like Luidia Ciliaris, members of the Solasteridae have ten to fifteen arms while the Antarctic Labidiaster Annulatus can have up to fifty. Our brightly colorful appearance and symmetric shape of our body is making us value for design of habitants of Earth and for our art's fishes. Our appearance played one role in literature of habitant of planet. their legend and their culture.

The Water Vascular system of the starfish is a hydraulic System that its network is of fluid – filled canals and is concerned with locomotion, adhesion, food manipulation and gas exchange. This hydraulic system is operating about sending Oxygen and sending out Carbon dioxide that is value point for Science too. Oxygen that is dissolved on the water is distributed mainly on the body by fluid in the main activity while circulatory system may also play a minor role. Our Star fishes are moving so slowly just 15 cm or 6 in in minute like Leather Star or Dermasterias imbricate, while some others like Sand Star fish or Luidia Fliolata can travel at a speed of 2. 8 m or 9 Ft 2 in per minute.

Our longer life is depended by different species for example Leptasterias Hexactis has adult with weight 20 g or 0.7 oz, reaches sexual maturity in two years and lives for about ten years while Pisaster Ochraceus release so many eggs on sea every year, and has adult with weight 80 g or 2.8 oz, while it reaches maturity in five years and has maximum life span or longer until to 34 years.

Star fishes are preyed by other starfishes species, also tritons, crabs. gulls, and sea otters. Our beautiful star fishes are getting protecting by producing one toxin in their chemical armour body like tetrodotoxinor saponins in their body walls which have unpleasant flavours. Also we are suffering in our life from Vibrio Bacteria. So those are some elements for value of science today.

At that time the red Starfish started to introduce on stage all stars in line, Orange starfish that swim in front slowly with greetings, after Green starfish, behind was coming Purple star fish, that was following it Pink Star fish, after was swimming with greeting Blue starfish, at the end came with proud Royal Starfish two colors Bold Blue and orange aside. At that time all colorful free fishes gave one stronger applause.

Royal Starfish stood in front of the stage suddenly started its speech when fishes stopped their applauses.

I am Royal Starfish from Astropecten Articulatus's family. We are located in West Atlantic Ocean. Our family is called Astropectinidae also we located in one place of North America to East Coast also Mexican Gulf's place, through Caribbean paces to Brazil another place in South America Continent as habitants are naming it.

Our Royal Starfishes are feeding it selves with mollusks which it catches with its arms and then take to the mouth.

I have one problem to discuss here in this Science's convention because as all fishes heard here a lot information by our honor Red Star Fish that are so valuable for this Science's Convention.

Our Royal Star Fishes's family has some restriction rules that is making us more isolated and why we are so proud like Royal's family but we see last time that so many fishes are doing good about their living while are selling with good price their scales. We can't sale our heritage treasure because of our family's rule but we want to be part of Science's Studies and Art's job too, so we can forward like all others in our good living and so we can ensure the future of our younger generation, but not to destroy our family's culture and tradition.

Another big problem for us and all others Star fishes is that our number was reducing so much we were 50% of numbers that we were along time in our existence while now is tragic situation because our existence is in 5% compare with our numbers a long time ago, so we need help and protection in that way we will survive and will continue our existence. I am proud for my appearance two to three colors Blue or Purple to Orange and White, so beautiful and for that I am saying thank you to our honor God Poseidon but this is awaking jealousy to others water's animals so like for me and for all others Royal Starfishes we need help by this Science's Convention about protection and about participation of science that both those two factors will ensure our existence. I want to make remarkable that I like to study and to make experiment in Genetic of "Bio – Waters – World" in this way I will help my Royal Star fishes' community in their existence.

So I want to make my request in official way. Some fishes told me that this request or this application must to do with writing in corals but I am not sure about that so I need help.

Ooo sorry for my interrupting we are not using this way anymore interrupted Surgeon fish this Royal Star fish. All fishes will do request or application with voice, their voice radioactivity and frequencies through water's molecules will come to my place so I will take information about their request but for you maybe is more difficult because is new technology new method of communication, so if you have difficulties, you must to put your arms on sand after some sand of underwater you must to melt with your exchange hydraulic system with your oxygen and put those on Corals. I will come to see those like your sign of your arms 'print and I know that you have done your request for Science's experiment. After that I will call to talk with you.

Thank you very much doctor Surgeon Fish for your helping and explanation. I feel myself very grateful with your presence and your attention to me to my speech and my request. At that time Royal Star fish with full enthusiasm moved all this arms like greetings so did that action and other Red Star Fish and both said in same times like both were in resonance: Thank you very much all of you very beautiful colorful fishes for listening us. All fishes, started their strong applauses and cheering. Surgeon Fish with full smile said this is wonderful speech by both of them, very great information we learn so many interesting information plus we have now another application so another new member for our Science's group. Thank you to all of you for attending this very important Science of Convention.

Chapter 15

Yellow Seahorse Argue!

Was coming at the end of discussion by free fishes in this Science Convention, while Surgeon Fish was commenting some information by different free fishes, when one strong voice came from audience and all fishes turned their head to the direction where was coming this voice. All fishes saw between all others yellow group fishes one yellow beautiful Seahorse. At that time all fishes of audience was seeing this Seahorse, it started its speech while did not care for all. I am Yellow Seahorse, that I want to speak from here I do not care for that stage, because Surgeon Fish and all others leaders of this Science's convention did not care for me and my community.

Maybe they are jealous for our beautiful appearance, and most important because we are very, very proud because different from all of you we are swimming vertical with our head up not inside the water. So we are seeing straight up and no any fish can put us down. My name is Hippocampus Kuda also known as the common Seahorse, estuary seahorse, or yellow seahorse I am member of the family Syngnathidae seahorse and pipefishes, of the order Syngnathiformes, for my community. We are so important because of our character and our very good looking so many communities of free fishes and humans on Earth are creating so many legends for us.

For example in older time Greek And Romans' human believed the seahorse was an attribute of the sea God Poseidon / Neptune and more important was considered a symbol of strength and power, so we are strong and powerful. Another legend is given by European for Seahorse, these humans of this place that named Europe believed that Seahorse carries the souls of deceased sailors to the underworld, while is giving to them safe and protection until they met their soul's destination.

So we have and our very high level and good spiritual world and why some are naming us all seahorses vulnerable species. As you know we are swimming in vertical position. Our seahorse broods fertilized eggs in lower abdomen while our female are slightly smaller. Our incubation of eggs is 4-5 weeks, while supposed that fertilization is taking place when eggs are in pouch. Seahorses can be found to Indo – Pacific, including coral reefs. We are spending most of the time anchoring to coral reefs and branches with their tails, made necessary because we are poor swimmer.

We are living in temperature 72 grade, to 77 grade F, or 22 grade C to 25 Grade C. We never are tolerating above 80 grade or 27 grade C.

We can tolerate salinity from 18 part per thousand (ppt) to 36 ppt but deal for all seahorse is 32m ppt. Sea horses' food are only small animals because of their small mouth, like Brine shrimp and new born guppies rotifers, when other foods are unavailable Dophina is eaten too. Seahorses need to eat 4 – 5 times a day. We have so many value points, so many habitant in this earth planet loves us for our beauty for Suvenirs, for aquarium and for Chinese Medicine that is very important, Name of this place is China, but I am not telling to you this medicine that we are source of this I am keeping secret as all of you kept secret this Science's Convention from all of my Seahorses' community, but this medicine is so in need and is successful.

So now I am saying Good Bye and thank you for your listening of my speech, and why I decided with my persistence way to speak but you never invited me for that so have a wonderful Science's Convention. Seahorse did not left time to Surgeon Fish to finish its speech but it started its proud vertical swimming and was forwarding in its direction while he gave one strong skepticism to all others that started to give whistle and some to call it to stay. After That Surgeon fish got desperate while said to audience we all are not perfect maybe we did mistake and we lost one beautiful member also one strong medicine that is keeping secret by Seahorses. Anyway I am closing this Science's Convention today, while I am saying one big Thank You to all free fishes that gave to us one big information for Science and for our life, also we must to learn from our mistake and to give to all alternative to participate in this Science's Convention like was last case of beautiful Yellow Seahorse.

At that time Sharks started to yell and scream and were saying: Yes, yes, you need to give opportunity to all to speak in this science of convention. All fishes were listening and Surgeon Fish too, but it did not reply because after Seahorse, Surgeon fish did not want to complicate the situation and specific for those wilders because it was thinking more for security of all free fishes of this Science's Convention but it finished its speech in diplomatic way:

Yes we will work next time to make one big improvement about our very important Science's Convention. Thank You very much to all of you for attending this Convention. All free fishes gave one strong applause. At that time Surgeon Fish called some Yellow Chichlid Fishes to stay because it needs to talk together after this Science's Convention before Chichlid fishes to go to their place - home – water.

Chapter 16

Confuse Chichlid Yellow Fish

When Surgeon Fish finished its beautiful speech with a lot thank you to all fishes, colorful fishes started swimming in their home – water's direction while they were discussing in their group as they were by their community. Very enthusiast and nosily were talking purple fishes while they do not care that others fishes were getting wondering by their noise. One purple fish said that it liked so much the ending speech of honor leader Surgeon Fish and its apologize for what happened at last speech by Yellow Horse in this convention.

At that time Green Fishes interrupted its speech and started their comment, that this apologize was not specific for Yellow horse but was for all others things that must to improve in future this Science's Convention, in that way the Green Fishes wanted to hidden their jealousy for Yellow Horse. The Purple Fish replied that it understood that was for Yellow Horse, but maybe was and for so many things about organizing, because some community fishes have more favor to speak that others.

The Yellow Chichlid fishes, were swimming very calm and were listening this noise conversation. The beautiful Small Tropical Fish, made with sign to Silver Fish that not to involve in their conversation, while the small Chichlid Yellow fish asked with low voice:

I want to know why Surgeon Fish did not call to speak on stage the Yellow Horse? Is it member of wilder community's fishes?

We think that it is more worse than that, while Yellow Horse is so selfish but with its activity is damaging all others colorful fishes community while is putting all of them to fight with each other, so for that reason he did not get invite on stage of this Science's Convention, but what it is doing that is putting in war those community fishes?

Yellow fish thinks that it is handsome and is more strong and smart than others, so when it sees that some famous fishes are showing skepticism to it, with emergency and persistence is using other fishes some innocents but some are like it and is preparing scenario to fight others, so Surgeon Fish knows that situation but did not want to argue and said one formal apologize, but really it know very well Yellow Horse, that is it the reason.

I thought that Yellow Horse was good and is so handsome while it got hurt for non-invitation I never know about this situation.

Ooo Yes Yellow horse has good looking but is so selfish never is thinking for good for others only for those that will work in its interest, so this is not good for anyone free fishes'community, they are tired to fight they want peace and normal life under water that is it.

Very strange this situation said small Chichlid yellow fish, so strange for really one very big conflict is existing to good looking of this Yellow Horse and its strange brain for intrigues and its strange heart too. Really small Chichlid yellow fish did not want to say for Yellow Horse that had bad mind and bad heart, because anyway it is Yellow Horse so was in yellow community of free fishes.. So Silver fish and Tropical fish saw small yellow Chichlid fish and started to laugh while they said:

Of course you want to protect because is Yellow Horse and is in your yellow family fishes, but we must to accept facts as they are, and your darling good looking horse is like that what we are telling to you, while they were continuing to laugh, but small Chichlid Fish got confused but never spoke anymore but was swimming quiet while was listening its friends Silver fish and Tropical fish. They wondered for its quickly stopped speech and said at same time, why you are so quiet dear Chichlid yellow fish, are you mad?

No I am listening you.

Ooooo I thought you are desperate for Yellow Horse, said Silver fish?

Really I want to ask something, may I ask both of you?

Yes you can ask us said Silver Fish.

Are Surgeon fish and others jealous for Yellow Horse?

About what to be jealous for Yellow Horse for its good looking, we have a lot very, very beautiful free fishes, so it is not about that.

No, no, answered Chichlid small fish I am asking for its activity, about Yellow Horse business or its success?

But he never had any success in its business, what you are talking about? said in same time both fishes, It never was shining in its business but sometimes had failures.

But why Yellow Horse was speaking in capricious way?

Because so it is, needs attention by others. If Yellow Horse had success, absolutely it will be first on stage to speak in this Convention of Science.

Why this Yellow Horse is doing that behavior, why it is claiming for something that it does not deserve, I can't understand that situation said Chichlid yellow fish.

I am telling you now answered Tropical Fish. This behavior is result of some behavior that involve phenomenon of Halucinacione and specific I am explaining in detail it is:

Enthoptic Phenomena (archaeology) that is "ART and ability to comprehend it are more dependent on kinds of mental imagery and the ability to manipulate mental images than on intelligence". If this phenomenon this Yellow Horse will make reality will be so good, but it never did anything reality and so is thinking that its imaginary is fact, but nothing is fact. Really if it does not work hard to make reality, so it is suffering by this conflict imaginary and not doing fact, but is thinking that imaginary is everything. So without its success is fighting in wild way others for their accomplishment.

Now I understood said to Tropical Fish the Chichlid Yellow fish. Really now I understood, so it is doing only picture in its brain and is closing its eyes with this dreaming but never is working to make reality.

Yes that is it, and is something else when others have imaginary and hard work he is manipulating to make that job on its own that is so bad and is irritating others so is coming bad opinion and big conflict.

Yes you are right said at that time Chichlid fish and Silver fish too.

Okay now you know continued its speech Tropical fish, that this phenomenon is impacting our life and as we all know nothing is coming free by the sky without job. So is not only imaginary and reflect to our consciousness but it is to work very hard and to make reality this imaginary, that is not so easy for some like Yellow Horse, with speculation never is coming success.

It can have some current result with its speculation or mixer others to put them in war, but it is not going longer, to make imaginary, reality is a longer process that wants hard work in persistence way with full passion and patience. Yellow Horse likes only himself and more important it wants quickly profit of others' job but never is going to happen in longer process quickly profit, so it is failure. I understood it maybe wants scales because of some problems of financial situation but for that free fishes has one board of helping, so for big things is not this profit.

Thank you very much Dear Tropical fish for your Explanation said Chichlid Yellow fish.

At that time one Red couple fishes were arguing strongly while were swimming closely to Tropical, fish, Silver fish and Chichlid Fish. They were discussing very nervous about one Science matter, but their conversation interrupted Perch Chichlid fish that was swimming behind those two Red fishes.

You do not need to fight each other for your topic. One of red fish said: I do not want that my brother to give away our secret of our experiment about feeding Red fish, while he wants to leave our family and to go to work alone with some others fishes to Surgeon Fish.

Perch Fish saw one family's conflict while said: Listen me if this experiment is good for so many others free fishes you must to be tolerant and to help science.

Yes it is for good of all free fishes, but our family and I am working so longer so hard and we need to be our experiment, when we will have result we need our name to this science's experiment and its good result, after we will put in market. For that problem I am strict and I am not tolerating my dear capricious brother that wants to betray us only to show up itself. Really we need some times to finish this experiment and to take result, but my brother is connected with some adventurer fishes that want to make in their own our experiment, that never is going to happen, so finished its speech Red fish while was seeing straight to Perch Fish.

I do not know how your brother is betraying you, how is giving information of this experiment to others?

I am telling you in its presence, it has its girlfriend that is giving information, while this girlfriend is giving to others, while is promising to my brother great future with high promotion that is not true is fake, they want to take our experiment to steal our experiment and to give good name about science to those adventurers fishes.

At that time Perch Fish approved and said:

You are very right, is your hard work, your creative idea, we never will allow others fishes to steal good job of science from science fishes like you, you are very right, I wish for you big success and we need to see your result in future in Science's Convention.

Thank you very much Perch fish said Red fish while all of them were swimming together while changed topic because all fishes of this group that was created did not want to create conflict during this swimming after this successful Science's Convention.

Chapter 17

Funny Neon Pink Fishes' inspiration!

While all group of free fishes were swimming back to their water home's place one very happy group of Pink Fishes got attention of all others participants of this Science of Convention. They were doing noise different their characters, so all others colorful beautiful free fishes turned their head and were watching. Those Funny Neon Pink Fishes were discussing loudly about this Science's Convention while with full euphoria started to express their feelings and their suggestions for future. Strangely they were not quiet and calm, like before and for the first time in their existence they started to be noisy fishes that gave wondering to all other fishes. So some Blue fishes and Red Fishes too. that were swimming close with them started to speak more slowly and to listen what they said.

One couple Funny Neon Pink Fishes while was hugging each other with full happiness with their transparent arms, that looks were in full love, said loudly to others:

We will start our experiment about chemistry that will help us about our body color. His fiancé said full with optimism that it likes modern fashion, it likes make up cosmetic, also likes metamorphoses, so likes so much to work in science of chemistry and genetic too, for that reason it will start to study University of Biochemistry while will continue experiment with its lover about changing color. They were explaining to their others Funny Neon Pink Fishes about their experiment.

One Pink Fish asked them: What is yours experiment?

We have study how to change the color of Blood Parrot fish to pink color until too bright pink color, and we had a very good result, beyond our expectation, so we are very happy. As you know that Blood Parrot Fishes are hybrid so are some difficulties to work with them, also those fishes has different colors like mainly orange, also have red color, yellow gray, specific they have vertical mouth and sometimes those fishes can't close fully mouth. So we have worked very hard while we have written everything on our corals notebooks all result of experiment. We faced some difficulties because Blood Parrot Fishes are so selection fishes about nutrition and more about malnutrition, so was very big and hard job for us to find in big distance specific food – nutrition to continue our experiment. At the end we changed the color of Blood Parrot Fish from bold orange to very bright pink color. This is one very good or more exact great result about chemistry's, but we will work and for Genetic too. So this beautiful Funny Neon Pink Fish was explaining very enthusiast about their experiment but its male Funny Neon Pink Fish time by time was keeping its transparent arm in front of its fiancé fish's mouth like accidentally and was saying sorry time by time. The female Funny Neon Pink fish did not understand its action and time by time said to it what is wrong with you and your arm that are putting in front of my mouth? … but male fish did not speak. So its fiancé was continuing its speech very enthusiasts while other free fishes were listening. At that time one red Fish asked it: What you did that you changed the color of Blood Parrot:

Okay I am telling you just right now: We started to work with….:

…But at that time one strong yelling and screaming by its fiancé came through the clear blue water that scared all others small free fishes:

Is confidentiality of our experiment and our study what we worked with!

Right, right!

At that time its fiancé understood its action that time by time was putting its arms in front of female fish's mouth, so female fish stopped its speech. The male Funny Neon Pink fish continued: We will work very hard for good result so we will give so much opportunity so many females

Free fishes to change their color as they want or what those fishes likes. We will help cosmetic in our water world, chemistry's science, food industry of Water's World about nutrition and so many others economic fields in our water's world.

All other fishes gave one huge cheer and full enthusiasm, but behind this beautiful lovely couple Funny Neon Pink Fishes, spoke nervous one Blue Fish: Why you are not telling about your work.

It is my experiment until I to have full result I do not think to speak more, answered strict Funny Neon Pink Fish to Blue Fish. At that time another Blue Fish that saw during this conversation was coming one conflict tried to interrupt their conversation, but really Blue Fish continued with persistence its speech to blame beautiful Funny Neon Pink Fish. Really Blue fish became nervous and all its scales came up, when its Blue Fish friend saw its scales were lifting up above its body, he was swimming in front of it and said:

We need to swim in other direction for some reasons. At that time all other fishes stopped their enthusiasm and conversation while all saw that Blue Fish took its friend to her direction to swim, after that scenario Funny Neon Pink Fish said with high voice:

What strange Blue Fish, was forcing me to speak for my unfinished job. I am not operating in that way also it is not my boss.

Another beautiful Pink Fish said loudly:

Please dear Funny Neon Pink Fish do not care for its speech, because I have heard a lot about this stranger Blue Fish, he is forcing and others successful fishes to give their secret of their job and it is making like its own job, or its creative idea, so do not tell to it but continue your experiment, also I wish for you big success and we need your result as soon as possible because time by time we like to change for fun and fashion our color but really we are very proud for our very beautiful pink color that our honor Water's King Poseidon gave to us, we can say one big thank you to our God Poseidon, but time by time we need to change for hobby. So continue your experiment.

All others fishes said loudly with full happiness:

Yes you must to continue your experiment and to give us your good result, so Congratulation !

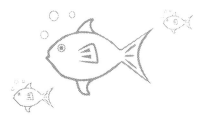

Chapter 18

Turtles' Conversation!

Behind those very beautiful colorful group of fishes were swimming so slowly but quiet the Big Turtle while around it were some small young turtles and its body guards fishes. The Big Turtle said in sweet way but very determinant to small young turtles:

Did you hear these couples of red fishes what they were talking and were thinking about their experiment and their secret to keep by others, so we will do. First of all when we will arrive to our water – home place we need to call all our turtles' family that are living close to our area and to make one meeting.

The young small turtles approved and said in same time yes we must to do that because we have so many value in our body about science of water's animals.

Yes we have so many value repeated big turtle while look that it was thinking so serious. We should do it.

As you know we have one very specific point on our build body about, that our shell can function as PH buffer. During winter period time such trapped beneath ice or within anoxic mud at the bottom of ponds, to endure through anoxic conditions, our shell utilize two general conditions mechanism while releases carbonate buffers and uptakes lactic acid. This phenomenon is happening during winter period time.

Really we have so many jobs to do about our improvement because as you know our shell has good and bad points about our life, so we need to make experiment how we can make more fast our reflexes because we are so low about that because our shell is preventing us to make our fast reflection.

We are proud for our long term memory until to 7, 5 months but we need to work for that.

One youngest small turtles reminded to Big Turtle: Dear Big honor Turtle I want to say something, may I say?

Yes you can say we all Turtles' family needs to hear suggestions and creative ideas or novation by our youngest Turtles. So what is your idea?

We can do one specific experiment to help all others fishes or reptiles of water and some that are water and ground's living like us that have built body like us or close with us too, about salt that are taking by water. As you know we have glands near to our eyes that produce salty tears that rid our bodies of excess salt taken in from the water that we drink. So if we are doing one experiment how we can create one this mechanism to others fishes we can help them to send out salt of very salt water so their body will get not sick.

Ooooo this is the best creative idea that I am listening so with this experiment that will be one big discovering or creative job in our water – animals' science we will help so many free fishes about this phenomenon so we will put our name in golden honor book of water's science.

At that time other youngest turtle said: I have another idea too, may I speak dear honor big turtle?

Yes you can say but what is about your idea dear youngest turtle?... asked the big turtle.

Okay I think we need to work about our vision because our vision has so much restriction. We all know that are some our Turtles in our family that most of them are spending their lives on land while have their eyes down at objects in front of them, so their vision is restricted, so we need to work about the vision of those kind of Turtles. I think we need to change something in their retina or cornea of their eyes or to change something in their intercells of eye's liquid so their vision to open more around, because we can't change their morphologic of eye.

Other youngest Turtle interrupted so fast other Turtle speech and said, I am very agree with your very great suggestions but I want to make one addition we can change their eye morphologic that some Turtles has their eyes very close to the top head, by genetic, we can make some our Turtles to make their eggs with some other animals so they can improve their morphologic of their eye!

So you think that some members of our family to change our tradition and to create relationship with others families and to take their material genetic?

Yes why not if this is for good of Turtle why not?

This is the most interesting idea but we will face so many obstacles because we need to work with youngest Turtles to give new direction about their love but their parents can prevent those because of their tradition. So this phenomenon is taking time but we need to do that but step by step because we do not need their parents to say that we are changing their race or their family's body's morphologic, because at least and last we do not need war but we need to make science. Yes you are right said at same time two youngest Turtles that were so excited by this Science's Convention and got so much inspiration for future.

After that creative idea big Turtle was explaining slowly, slowly but very quiet to youngest small Turtle while all others Chichlid yellow fishes its guardians were listening with full interest. Really our water's Science fishes and Earth's habitants are working in genetic science and have some good result by their researches about examining the Turtles genome for longevity genes. They have discover a Turtle's organs do no gradually break down or become less efficient over time, unlike other animals. It was found by them that liver, lungs and kidneys of centenarian Turtle are virtually indistinguishable from those of its immature counterpart, while this group of Turtle is known as a Bale.

All its guardians Chichlid fishes and youngest Turtles were thinking: How smart is this Big honor Turtle, how it knows so many things about Science that is explaining to us, so we need to study like her. At that time two youngest small Turtles whispered to each other, we need to study like our big honor Turtle, because we never knew before how many value points we have in our body for Water's animals' science.

The Big honor Turtle said with enthusiasm:

We will do what we have discussing now but I have another proposition that I will say to this our future meeting of Turtles' family.

I think that our experiment to put our code name" Chelonian", because this name really is well known name for us by scientists, veterinarians and all conservations working habitants of Earth. As you know this name came by one very interesting place of our Planet that named Helenic and their language is Greek language, so this word or this name means "Tortoise" that is calling in sweet way us with love feelings.

Ooo this is wonderful name to call our experiment approved others youngest Turtle while all others Chichlid Yellow fishes said this is so good to happen while where swimming with full care and were watching all around to protect big honor Turtle. Chichlid guardians Yellow Fishes were watching all around and were enjoying all around blue water that looks like got one very beautiful decoration by so many colorful and beautiful free fishes that were participants of this Science's Convention. Really this view looks like one very longer and beautiful pilgrimage by colorful fishes.

Chapter 19

Shark's Conversation!

During their swimming on blue water all free fishes that were attending that Science of Convention suddenly heard one strong noise while the water looks that wanted to meet the white cloud up to the sky so high was going water like water fall that was lifting up blue but was coming down like white water or white waves. All free fishes got wondering. All turned their head and saw on small crew of sharks that were swimming very fast and were speaking so loudly.

One shark was yelling to their leaders: Why you did not speak in that Convention why you did not allow us to interrupt them because they showed up big skepticism for us. We have big value too.

At that time one yellow Chichlid fish interrupted that speech of this youngest shark while said:

What value have you Sharks?.., tell us, all around this planet knows you with one name like Predators of others fishes and Earth habitants.

I am telling you that this concept is relative is not for all of us also I am telling to all of you all ours value points. Chichlid yellow fish was continuing with persistence way its speech: Nothing you can tell us because all of you sharks are Predators, no one has done selection in your family for peace sharks in water blue's world.

It is not true I am telling you now leave me and give time to speak and why Surgeon fish never allowed us to speak in that Science's Convention.

At that time two Red fishes made sign with their arms to Chichlid yellow fish to allow shark to speak, while were lifting their scales up, that was sign of being alert for unexpected situation. At that time all fishes slow down their swimming and they created one circle while they wanted to listen the capricious shark.

At that time the youngest Shark that looked that wanted attention and was preparing itself for big career in water's world, started its speech:

In our shark family that we are well known in water blue's world with name Selachimorpha or short Selachii, contrary to habitants belief and to all free fishes, only a few sharks are dangerous to habitants of Earth and some free fishes. We have more than 470 species, only four have been involved in significant number of fatal, like unprovoked attacks on habitants of earth or humans as they are calling themselves., Those dangers Sharks, are the Great white Sharks, Oceanic Whitetip Sharks, Tiger Sharks and Bull Sharks, these are very big, large powerful predators and so many times attack and kill water's habitants or humans of Earth planet, but as you know all others families has good and bad members in their circles.

But what you want to say for event of Jersey shore Shark attack 1916 one place of Earth planet that we have read in our water blue' Encyclopedia? asked quickly the Chichlid yellow fish,

Of course those dangerous large Sharks are doing some times unprovoked attack and are killing humans of this Earth planet but I am telling you something that Science's fishes to give news with their writing to corals or with Turtles that are going from water to ground, this news but sorry I am betraying in relative way my family so my advice and secret is for these Habitants of Earth or humans:

To help avoid an unprovoked attack, habitants of Earth planet or Humans as they are calling themselves, should not wear jewelry or metal that is shiny and refrain from splashing around too much because sharks have some specific in their retina of eyes that I will tell now so those humans will attract Sharks in their direction.

I think we do not need your advice for helping habitants of Earth planet because you never can change my mind for yours wilders nature, of all of you Sharks. What you will say to me if I am saying some numbers by Statistic's science of humans of this Earth that we have got in our way and have written to our Encyclopedia's book. Those numbers are that in

2006 the big studied, The International Shark Attack File (ISAF) as have named the humans of planet their studied' s material that undertook an investigation into 96 alleged Sharks attacks, confirmed 62 of them as unprovoked and 16 as provoked also the number of fatalities worldwide per year between 2001 and 2006 from unprovoked shark attacks is 4, 3, so what you will say now, all of you are dangerous in this planet's life.

Ouuu so you are saying, you Chichchchchclid Yellow Fish, started to make irony the capricious Shark.

Yes so I am saying that all of you are wilder and danger Sharks.

Okay are you allowing me to finish my speech or not? asked capricious the youngest Shark while was lifting with its tale water up to show that it does not care for Yellow fish.

Okay I am allowing you to finish your speech with your value points about science but this does not influence to change our opinion about your dangerous character..

Okay I am finishing my speech said youngest shark while all others free fishes were listening it.

As you all know that Sharks'eyes are similar to the eyes of other vertebrates, so Sharks have similar lenses, corneas and retinas, so though their eyesight is well adapted to the marine environment with help of tissue called Tapetum Lucidum. This Tissue is behind the retina and reflect light back to it, so that is increasing visibility in dark water of sharks. This is big point with high value for Science's Convention and for Ocular science too.

Some Sharks have membrane that is covering their eyes and some does not have membrane to cover their eyes. Some of our Sharks has colorblind, some are watching so good gray and green color some they do not see color object. This phenomenon of sight has to do with electro and chemoreceptions activity.

About hearing our sharks has a small opening on each side of their heads (not the spiracle) leads to the inner ear through a thin channel. The lateral line shows a similar arrangement, and is open to the environment via a series of opening that called lateral line pores.

These tow vibration - and sounds – detecting organs that are grouped together as the acoustic – lateralis system. I am telling to all of you that to bony fishes and tetrapods the external opening into the inner ear has been lost. So it is hard to test the hearing of Sharks they may have a sharp sense of hearing and can possibly hear prey from many miles away.

Suddenly at that time the big Turtle spoke slowly but very determinant: Please you Shark more better of you to tell us about your teeth, that you are creating so fatalities all around.

Oooo you honor Big Turtle, I am understanding your irony and maybe is more high, is sarcasm about our teeth but and about our teeth we have something good for Science's Convention and for Water Blue's Science, while it was moving full with pleasure its tale because the "Youngest Shark" understood that it got attention, for that he was fighting to get attention, because he knew very well that can't change their opinion about wilder nature of sharks.

Sharks all the time constantly are replacing throughout life their teeth. The rate or tooth replacement varies from once 8 to 10 days to several months. Through their life some sharks lose 30,000 or more teeth in their life time. In most species teeth are replaced one at time as opposed to the stimultuous replacement of an entire row, which is observed in Cookiecutter shark.

Tooth shape depends on the Shark's diet, for example those that feed on mollusks and crustaceans have dense and flattened teeth used for crushing, those that feed on fish have needle – like teeth for gripping, while those that feed on larger prey such a mammals have lower teeth for gripping and triangular upper teeth with serrated edges for cutting. While the teeth of plankton – feeders such as the basking Shark are small and no functional.

Our Sharks' family members are found in all seas.

This is tragedy of this planet and water blue's world interrupted the Chichlid yellow fish. But I do not want to prejudice the job of our Honor god Poseidon finished its phrase Chichlid Yellow Fish.

Ooo really replied Shark, why only you are in need for this "Water blue's World"?

No, no, no, because our honor god Poseidon created all of you together with us to tell this "Water Blue's World" "The United of Opposite Theory" that good and bad are living together, or soft and wild, so Poseidon created you wilders Sharks only to tell what is opposite of good that is you the bad or worse.

We are so proud because we are selective continued the youngest Shark while was showing that did not care for irony and blaming of Yellow Chichlid Fish, we are not living or swimming in fresh water only Bull Sharks and River Sharks can swim both in seawater and fresh water. Sharks are common down of 2000 meter in water (7,000ft) and some live even deeper but they almost entirely absent below 3, 000 metres (10,000 ft). the deepest report shark in our Sharks' Encyclopedia's book confirmed of a Shark a Portuguese Dog Fish (name originated by one beautiful place close to Atlantic Ocean in this planet) at 3,700 metres (12,000 ft).

I have another big news for this Science's Convention and for "Water blue's World" about Electroreception. To our Sharks' bodies are Ampullae of Lorenzini are the electroreceptor organs that are in hundreds or thousands in numbers. Sharks use Ampullae of Lorenzini to detect the electromagnetic fields that all living things produce. This helps Sharks to find prey specific Hammerhead Shark. The Sharks have the greatest electrical sensitivity of any animal. Sharks find prey hidden in sand by detecting the electric fields they produce. Ocean Current moving in the magnetic field of the Earth also generate electric fields that Sharks can use for orientation and possibly navigation.

So this phenomenon that we named it Electroreception is so value for Water Science and is one big news for Science's Convention. All were listening carefully but no one wanted to give value for that point that really had big value for science, but Chichlid Yellow Fish said: And why is value this point this point can't compensate your wilder nature in our water kingdom. But I want to ask you something.

May I ask you?.. said Chichlid Beautiful Yellow fish.

Yes you can.

What you have to do with beautiful elegant Anguilla Rostrate?

Simple I am swimming with it around. Answered Youngest Capricious Shark

But I do not like you to swim with Anguilla Rostrate, it is soft you are wilder.

I am not wilder.

You are wilder and danger.

I am not danger. Also I think Anguilla Rostrate loves me?

What? Pspspspsp started to make Chichlid Yellow fish, Anguilla Rostrate loves you?!,... never is true maybe is scared by you and is coming around to make you soft because you are swimming all around its' place and wants to keep in safe its Anguilla's community. I know so many things about you.

What you have to do with my private life, ooo Chichlid Yellow fish? This is not matter of Science's Convention replied with sarcasm Youngest Shark, because I came here for Science not to tell you about romantic story.

Who will have romantic story you and all of you Sharks,??!! Ooo my honor Poseidon, what you have created those wilder, but how do you know about romantic story and love, you are losing 30, 000 teeth during yours life all of you Sharks because you are killers and you are damaging your teeth I do not believe that your teeth are getting damages by "Love and Kisses". You do not know what is "Love"!, For that I am asking about beautiful Anguilla Rostrate, it is innocent and looks quiet and shy.

Anguilla loves me started to scream and yell the Youngest Shark.! This is my private life my problem not your. You Chichlid better to see your life.

Yes I am seeing my life and others free fishes life to make this water "Blue Water's World" better place for all of us fishes, so I and all of us need to protect Anguilla Rostrate by you wilder while for those value points of Sharks' bodies will decide board and Surgeon Fish not me but I will speak up with all my friends, so thank you very much for your speech but not thank you for swimming with Anguilla Rostrate because I and all of us will not allow you to play danger with Anguilla. We do not believe you, so continue your swimming with your friends Sharks too. All others Sharks that were behind the Youngest Shark spoke with one voice:

Of course we will swim together because we are all famous Sharks.

Famous in your brain in your head with so many pores as explained this Shark but not in our brain and our head, better of all of you to close some pores so some part danger of thinking never to come out of yours head.

At that time all others Sharks started to make different replies with irony: Ooo you know because you learnt so many things by this Science's Convention that was not Convention but parade of some of you to show up, but we do not need to show up because we are famous, while you are fighting to become famous all of you free fishes.

At that time two Red Fishes said in same time like in resonance:

Oooo Yes you are famous but isolated by all others "Water's World" because no one loves you so your famous's value is zero or has zero points. We do not need your famous but danger name also we do not need your value points because we will do experiment and we will discover new way of electroreception and about chemistry of your eyes's retina. So good luck in your wilders community, closed their speech the beautiful Red Fishes. Sharks did not speak but they started to swim and one of them spoke loudly, we will come to others Science's Convention we will come.

Chapter 20

Marine Fishes' Conversation

After Shark speech during swimming also after big noise that created those Sharks that were claiming for attendance for next future Science's Convention, one very beautiful marine fishes' s crew appeared. Marine fishes that were so white and like transparent like clear water were swimming full with happiness between other beautiful colorful fishes. Really this transparent fishes' crew got big attention by all others and they got wondering how they did not see those in Science's Convention. All others fishes were watching those very beautiful fishes, with their elegant transparent cover by their luxury scales, while they looked so peaceful. The Red couple Fishes asked Chichlid Yellow fish: Do you know those fishes, because we never saw them in our life.

Chichlid Yellow fish answered:

Yes I know this fishes' community. It is Marine fishes, I have some information about this community I will tell you right now. There are thousands of species of Marine Fishes but dominant are tiny Zooplankton to enormous Whales that are with different adaption in water's world.

Marine fishes are drinking saltwater but are eliminating that salt by their gills. They are naming that phenomenon Salt regulation.

Marine fishes and other organisms that are living underwater are taking oxygen from water also through their gills and skin. It is specific about their Marine Marmmals because are coming to water surface to breathe. This is the reason that the deep – diving Whales have blowholes on top of their head, so they can breathe on surface of water while their body is underwater.

Whales can stay underwater without breathing for one hour because of efficiency of their lungs that during their breathing whales are exchanging up to 90$ of their lungs volume with each breath. At that process of breathing Whales are ensuring high volume of amount of oxygen in their blood and muscles when are diving. They are naming that phenomenon Oxygen's system circle. This is really one process physiologic that we need to study in our Water's Science or in our future Science's Convention. We have some information that some science's habitants of the Earth, or humans as they are naming themselves are studying that phenomenon and sometimes they have created some water machines that are using that way to swim underwater, but I do not have so much specific information about that because they are so secretive, but sometimes I saw those stranger machines underwater, maybe they studied those Marine Fishes and created their machines who knows.

The most interesting point of Marine fishes compare with other water animals or ocean that have cold - blooded (ectothermic) is science term for that and their temperature is same like environment that are living, Marine Mammals have warm – blooded (endothermic) is science term too, meaning that they need to keep their internal body temperature constant no matter the water temperature is. For that they have special consideration. Marine Mammals have layer of blubber under the skin that allows them to keep their internal body temperature warm even in cold Ocean. Some Artic Marine Mammal and bowhead Whales, has two feet thick this blubber layer.

Do you know that in our Ocean pressure increases 15 pound per square inch for every 33 feet of water. Some information are telling us that some Ocean's animals do not change water depth like Whales but sea' Turtles and seals sometimes travel from shallow water to great depth water several times in a single days, so this is big issue while we need to study but I do not want now that honor turtle to hear me what I am saying this is big discovery about their build body how they are affording this pressure of water? How can they do it? This is another problem that need solution for our water science, this is another thing to discover about water pressure in our Ocean and need Science's study.

About light so many corals reef and their associated algae are living in shallow, Clearwater that can be easy to penetrate the sun light but some water animals. Some fishes of this community in depth of Ocean lost their eyes or their pigmentation because are not necessary. Some others organisms are bioluminescent, because they are using light –giving bacteria or their own light- producing organs to attract prey.

Whales has n another specific phenomenon, they locate their prey by using echolocation and their hearing since underwater visibility and light level can change time by time.

Those marines in the intertidal zone do not need to deal with high water pressure but they need to withstand with the high of wind and waves' pressure. Many Marine invertebrates and plants in that habitat have the ability to cling or attachment their body strong to the rock so they are not washed away also they have hard shells for protection.

At that time one Red fish asked with full curiosity the Chichlid yellow Fish: How do you know all these information about those marine fishes.

Ooo easy way all the time when my parents are going to their friends or somewhere, I am reading in hidden way their writing or different notice and sometimes I am swimming to Enciclopedia's Coral book and I read over there.

Ooo this the great job that you are doing. I want to study like you, can you tell me where is this

Enciclopedia's Coral Book?

Yes I will tell you answered with full happiness the Chichlid Yellow fish, one day that I will be free I will come to take you and to go together to that place.

Thank you very much dear Chichlid Yellow fish said with smile and full happiness red fish.

So you know so much about Marine fishes now but i need to call the beautiful Spectacle Parrot fish to get full information about those transparent fishes. At that time Chichlid yellow fish started its whistle, phs,phs,phs after some minutes all of them saw one beautiful fish with color like ocean's water this was spectacle Parrot Fish.

When it was staying in front of them Chichlid said to it: I need information about Marine Fish. Suddenly the Spectacle Parrot fish started to say quickly.

Common white fish with Coegonus lavaretus in genus Coregonus.

Please do not rush Parrot fish I need information about those beautiful white transparent white fishes.

Okay they are Transparent white fishes but and some other communities have pink color. Their name is Parambassis Ranga but so many are calling them all around water world like Indian glassy fish, or Indian Glassy perch, or Indian X-ray fish. Their bonds looks by outside are so beautiful and are coming from Ambassidae's family.

Their name India is originated by one place on earth planet also their locate living is another place named Asia by Humans of earth. Those Fishes are so beautiful and so many are using for their beauty in some glass aquarium as are naming by Earth habitants. Because of their beauty Indiana Glassy fishes have so many romantic stories, you can see now, but all the time one Angel fish is protecting them.

White Betta Fish is in love with one of those transparent fishes and is a long time that is coming and swimming around this are but its rivaled by Pet Betta fish.

You will see right now they are close with us.

So all of them wanted to erase their curiosity and started to swim faster. After some minutes they saw one very beautiful transparent white fish that was swimming close with White Beta Fish but other like dark blue color that was Pet Betta Fish was swimming all around without control and anxious.

So all understood the situation, as explained to them the beautiful Parrot fish. At that time they heard the Indiana Glassy conversation with very beautiful white Betta fish., while they did with sign with their scales to each other for not noise.

Indiana Glassy while was swimming so happy said to White Betta Fish: it is a long time that you are swimming in my area and are disturbing my thoughts. I want to resist you persistence I tried but I can't resist anymore is not because of beauty appearance but because of you ggood heart and thoughts, so I am thinking every time for you now, but I have problem with my family and my community.

Why you have problem with your family and your community.

They have nothing to say about my appearance and my education.

No, no but they want for me to have relationship only with our transparent fishes's community, only this is reason. They are claiming for supremacy of our family because they are claiming for our beauty more than others.

What is this? And my family is so beauty with a big value, plus I am in love with you and I will convince mind of your parents and your friends too, also I will take you to see my laboratory of Science's Chemistry and why you are so fragile in your body, but I will create security that not any one chemical liquid to damage your beauty.

Really you will take me to your labor chemistry Science's place in laboratory?

Really I will take you over there. I need you to study for Chemistry's Science and to work all the time with me so we will be together every time everywhere.

Ooo this is sounding so good. I will work out about this matter with my family and I will give you answer soon.

So good thank you very much dear very beautiful Indiana Glassy for your good words and big hope to me about you, also thank you very much that you allow me to get into your life too.

I want to give you one advice White Betta fish!

What is this advice?

I am telling you that Alby fish is hired by my family like investigator about my daily activity, because they do not want for me to happen any bad thing, but you just to know that Alby fish is not so friendly is more wild than soft plus is so fast swimmer and is dedicate to its duty, I do not want you to create any conflict with this investigator fish so be careful until our problem to get good way.

Thank you very much for your advice I will take care for myself during my swimming said White Betta Fish while it never can cover his happiness and full smile. At that time the dark blue color Pet Betta Fish was swimming wild and was lifting in irregular way some water up by its speed, so all understood that was coming one time of conflict. At that time Chichlid Yellow fish to interrupt this time conflict started to swim with high speed and went close to Indiana Glassy and White Betta fish while hit them like accidentally, at same time said with low voice to White Betta Fish:

You need to go away now. Do it go away. White Betta Fish that understood situation gave one smile and one bye expression to Indiana Glassy and started swimming while said I see you later. Indiana Glassy was confuse about this interruption and got wondering by suddenly strange behavior of White Betta Fish, while was going to be together with its other White Transparent Fishes.

After that Chichilid Yellow fish was swimming to its group to Red Fishes and Spectacle Parrot Fish. All they said to Chichlid Yellow Fish: You did one wonderful job, while all of us are very happy for their love story and their good understanding with each other plus White Betta Fish is Science's fish and so beautiful. Thank you Chichlid Yellow Fish for your good action.

It is my duty to service to good fishes plus I think is time that all of us to work for Peace keeping in our Water's World.

At that time in front of them was swimming so fast the most beautiful fish with very characteristic shape and its tale, with white color dominated all over body but head with orange color and arms and its tale like decoration blue color, this was Angel Fish, that was swimming to arrive the white transparent Fishes' crew.

All were happy while were swimming for colorful view that was creating by fishes in ocean blue water.

Chapter 21

Yellow Chichlid fish's Family Party

During swimming to back home through the big beautiful crew of yellow Chichlid Fishes, the small Chichlid Fish with its Tropical fish started to discuss with enthusiasm about this Convention and about so many beautiful things that they saw over there. They knew that they were very, very, young for Science's Convention because they did not start their elementary school yet, but strangely they started to speak about their big new dream.

At that time small Chichlid Yellow fish' parents called it and ordered to swim with high speed to their water - home, when to arrive over there it must to call all its relatives and some friends that they said their names, to come to their water – home because of one big party that they will do. Small Yellow Chichlid fish gave one strong phssssssssphsssss with full happiness, while left Tropical Fish behind with its invitation for party and started to swim very fast. Yellow Chichlid Fish looked like was dancing during its swimming, that all understood that it was so happy.

While Yellow Chichlid Fish was swimming to its water – home, all other colorful fishes were swimming with laughing and talking while were saying for their new dreams and project for future and for their goals that they wanted to achieve until to come the time of new Science's Convention.

During its swimming the small Yellow Chichlid Fish saw that color of ocean looked more deep blue while air fresh was coming by wind that was pushing white waves on surface to huge the legs of white pelicans. Really the small Yellow Chichlid Fish was getting wondering by this white waves and White Pelicans,that this color was coming in intercalary way through blue water, while this situation gave one big pleasure to small Yellow Chichlid Fish. Yellow Fish started to laugh when saw white waves were touching and blocking the legs of Pelicans. During this trick game by white waves the Pelicans tried to move and fly above water but they can't because of capricious white waves. Really was big inspiration this scenario through blue water were coming the sunrays while were making bright inside water. Through this blue and bright light small Chichlid Fish was enjoying the game of white waves with legs of white pelicans. Everything looked in harmony while water's God Poseidon was painting all this water's surface with full happiness to give more high level of pleasure of Small Yellow Chichlid Fish. At that time Yellow Chihclid Fish spoke with high voice:

OOOO my God how beautiful is in our water's world.

During its swimming the Yellow Chichlid Fish was thinking how to organize gathering of all his relatives, friends and others. When it arrives to its water – home gave signal to others fishes, to come close to it. After sometimes came from all around so many Yellow Chichlid Fishes to see Yellow Fish that came from Science's Convention

Yellow Chichlid Fish started its speech to all those free fishes that came close to it. I called you to come here only for this treason:

My parents were with me in Science's Convention of Free Fishes, after this Convention they decided to organize one big party to honor this Science's Convention.

I want to give to all of you my relatives and friends of mine big pleasure with my invitation, another news is, that in this big party will come some famous beautiful and colorful intellectual and simple free fishes, so I need by all of you to show very good behavior and yours attitude to be classy. Really this is not advice by my parents but this is from my heart because you are my relatives and my friends so I want that you to give one very good impression to our new guests.

One youngest yellow fish asked immediately: Okay I understood your preoccupation for our behavior and our attitude, but really we do not know your guests if they are really classy too, because their attendance in Science of Convention is not proof or evidence of their classy fishes.

I did not say that they are classy our guests, but I said I need my relatives and my friends to be classy in this big party. While about them, really they are coming from different waters – are with different professions and level of education and some are simple with high school education but they like science are reading so many books and magazines to corals under water so they are claiming in future to became Science's fishes. You are my relatives and friends, so I am preparing you for this party. I need to defend you by any misunderstanding that will send to any unexpected conflict. So in short way I am warning you, protecting you and defending you with all my vast God given powers.

While the Yellow Chichlid Fish was working for organizing of this big party in its water – home, his parents were swimming more slowly with some their other free fishes friends and were discussing about this big Convention that they attended. They said that this Science's Convention was so important, about progressing of Research of Science, plus was so wonderful over there because they saw so many others fishes with their beauty and so colorful with different shape of their bodies, that they never knew that were existing. We are enthusiast to make this big party to honor this big Science's Convention, but we are not sure how is doing preparing situation our dear youngest Chichlid Yellow Fish with his friends about participation. After some times of swimming Parents of Yellow Chichlid Fish arrived to their water – home with some of their preferred guests – friends, free fishes. They got big surprise when they saw so many yellow fishes, some white octopod fish also some yellow, red pink and blue stars, and some other colorful fishes, also some youngest small beautiful Turtles.. They shocked when they saw that its youngest beautiful Yellow Chihclid Fish son has prepared with its close friends one very beautiful big thick ice Table and above its were so many corals like decoration leaves and grass that they took in deep underwater. Its Yellow Chichlid mother said loudly:

This is magnificence ice table that i saw in my entire life. Ooo my Water God Poseiond this is so beautiful. After it asked its son- fish: My dear small yellow Chichlid fish – son tell me who helped you for that or who gave this beautiful idea. Small Yellow Chichlid Fish answered quickly without taking breath while was putting in danger itself because water was going so fast through its mouth and was blocking its breath. I did, it is my creative idea, I have read before some times to one book of Corals about design and decoration, when I went to meet one friend of mine other yellow fish of other community.

Its parents really shocked by this new creative idea of their fish – son, they were so happy. They saw on table so many good foods like so many mollusks, some different small and big algae, they shocked, but its son explained to them that its friends helped to find all these foods. When they saw some other foods that never saw before, they got wondering. What is this delicious food? They asked after they tested.

The Yellow Small Chichlid Fish smiled and said this is secret, but I am telling to you my dear mother only to your ear like whisper. Its father fish laughed and said tell to your dear mother. So Small Yellow Chichlid fish swims close to its mother and started its whispered, because it did not want other free fishes to listen its secret.

Dear Mother, I am telling you that this delicious food for your big beautiful party brought for me one very beautiful American Chichlid Fish that has name of one very beautiful place of Earth planet also its lover Long Finned Chichlid Fish. So many times they are going to see in hidden way their friends that are taken by Habitants of Earth planet or humans as they are naming themselves. Their friends are isolated in some glass tanks that are naming by humans Aquarium. So this couple is creating one Organization to give freedom to their friends that are living isolated in Aquarium but this is taking some times. They have everything in those aquarium but they are not happy because are isolated, so are alone so they needs freedom, so this beautiful couple went to one of their friend and he sent out of this aquarium this delicious food that Humans are preparing for it. This food is prepared with Science about mixer and ingredient, so we need to test and to produce in future in our community. I have put some others from this food in hidden way only to study the ingredient of this food.

Oooo My water God Poseidon this is the great news that I have heard in my entire life. I never knew that my small Yellow Chichlid Fish – son is so smart and its preparing itself for science.. Thank you my dear Fish - son. While Its father said:

I am very proud for you my dear fish son.

At that time its mother asked again small Chichlid Yellow Fish: Dear son but how those free fishes went to Aquarium?

Dear mother, those fishes never went with their wishes, to Aquarium, they are so beautiful so their beauty has sometimes good and bad price, so some Earth Habitants or humans as they are calling themselves has some big value money that are doing business as we are doing with our scales. So they are thinking that with their big amount of those value monetary they can do what they want, also they are cooperating with some speculators humans that are working with fishes on blue water and are selling those colorful beautiful fishes and are isolating. They are claiming that they have created so wonderful condition for those free colorful fishes, but really fishes are isolated. Fishes wants to live free like us, so is big process now by this couple and their organization that needs more times and hard work to give freedom to them. Thank you very much my dear son for your explanation. I love you.

I love you my dear mother spoke with full happiness small Yellow Chichlid Fish. At that time all free colorful fishes came around the ice table to enjoy this party. The small Yellow Chichlid Fish's father stood in front of all other guests, while he smiled started its speech of welcome. While it was seeing all around this ice - table Yellow Peacock Chichlid Thunbnail Fish.

Peixe Chichledeo Africano autonocara Apa or Apache Peavrcock Chichlid Fish as called its fishes, friends,

Yellow Parrot Chichlid Fishes that were in big number in this party and looked very happy most of them were very younger,

The Yellow Chichlid Fish father of small Yellow Chihclid fish thought, how beautiful are all of them and how optimism have in their eyes and their smile, all are so beautiful and energize so we must to be proud for them.. So it started its speech:

It is my honor to welcome all of you in this party. We came from one very important historic Science's Convention with so many decisions for our life in water - blue's world, so we need now to enjoy our party for our big success and big result from this Science's Convention. I will help all of you youngest fishes and I will illuminate with glorious yours future. With our future job we will reveal miracle for our life. All of us will work together to embrace our future life and to find new wealth.

I personally will be so much devoted to all of you, to prepare all of you for new way of life. I will reach out for new friends but really I know and I believe that all of you are worth of my effort. So I am happy to have you and to be with you too.

At that time one huge sparkle of water came up from enthusiasm of all colorful fishes and their applauses. One group of Yellow Peacock Thunbnail started to divine in water so they created with their fluoreshent light one beautiful bouquet like fire fighter all around that created so much enthusiasm. At that time dancing in water started American Chihclid Fish with its lover Long Finned Chihclid Fish while was saying that will work for freedom of Aquarium's Fishes. After this couple in line came two by two some other couples of fishes, that were swimming – dancing with full happiness and different whistles psspsssspsss. Suddenly the Long Finned Chihclid Fish got by its arm Electric Blue Chihclid Fish and Dwarf Chichlid Fish and started to swim- dancing, after them came all others in line Yellow Chichlid fishes, Peix Chichlideo Africano Autonocara and all others colorful youngest fishes, So was created one longer line arm by arm of fishes that were dancing together while were creating some soft waves of blue water. This was the most beautiful view under water blue by swimmer - dancing of beautiful colorful free fishes through light of sun's rays that were coming with full energy by outside of water. Parents of small Yellow Chihclid fish, hugged each other and said to each other, we have so beautiful youngest colorful fishes,that are so energize. We have so much resource and potential for our future science, thanks the Water God Poseidon, for existence of our free fishes. We will have success in our research of Science of our life. We are happy for our free fishes, and we are so proud of them. We are with full hope for our bright future life. Now let's enjoy this beautiful party with full happiness.

At that time one crew of fishes grabbed both of them and put them in line of dancing with others. Sun's rays were playing with this beautiful colorful crew of fishes under water blue. The view was wonderful It looked like one colorful painting where blue water was decorated by so many colors of fishes. This view was with bright color that gave energize and optimism to all.

Free Fishes were enjoying their happy party for their new bright future life.

<u>End!</u>